Contents

Answers can be found online at hoddereducation.com/answers-and-extras

Working within the engineering and manufacturing sectors

Recall activities

1 Read your notes and textbook and then, from memory, jot down some ideas about what to include in a mind map that shows the seven types of manufacturing processes. Add as many **examples** of each as you can remember. Create your full mind map on a separate sheet of paper. After you have finished, go back to your notes and textbook and fill in anything you have forgotten.

> **Hint**
> Creating mind maps is an excellent way of organising information and helps recall.

....

....

....

....

....

2 Match the key terms to the definitions.

Term
Joining processes
Commissioning
Iterative design methodology
Cellular layout
One-off manufacturing
Design brief
Design for manufacture
Ergonomic design
COSHH
Operational envelope

Definition
A document which defines the purpose of a new product and outlines all the needs that must be met
A design approach that ensures a product is safe and comfortable to use by taking into consideration both human behaviour and the size, strength and limitations of the body
The area of three-dimensional space representing the maximum extent or reach of a robotic arm
Control of substances hazardous to health
Processes used to fix two or more components together into a larger assembly
The process of getting a newly installed machine or piece of equipment operational for handover to the customer
An approach to the design process, with structured opportunities to improve the design solution over several design cycles
Using skilled labour to produce a single product
An approach that ensures products incorporate features that support the use of a particular process in their manufacture
A manufacturing process that uses small, self-contained manufacturing cells for the batch production of variants within a family of similar components or products

Short-answer exam-style practice questions

1 There are six accepted descriptors for sustainable design. These are known as the six Rs. Here are five. What is the missing 'R'? (1 mark)

Reduce	**Refuse**	**Rethink**	**Reuse**	**Repurpose**

....

Hint

This is a learning and retention question. Revise facts like these to help gain individual marks that all add up. Try to remember the answer without looking it up, but if you really can't recall, go to the textbook and check. You will have a better chance of remembering next time.

2 What is the term given to the installation, commissioning and ongoing care of machines and equipment that ensures they are kept in good working order? (1 mark)

....

Hint

Again, this is a simple learning and retention question. There is just one mark, so only one answer is needed.

3 Explain some major differences between one-off and batch manufacturing. (3 marks)

....

....

....

Hint

There are three marks for this question, so you should give three good differences to gain full marks. These should be in sentence form, not one-word answers. Just putting 'cheaper' does not give enough information. 'The cost is less per item in batch manufacturing' is perfectly clear.

4 Why would an automotive manufacturer perform safety tests on their products? (4 marks)

....

....

....

....

Hint

This is a relatively high marking question, so take some time to answer it. Think of the various reasons for safety testing – for example, to conform to legislation, to protect the company's image and to sell more produce. As always, if there are four marks then you should try to include at least four points.

5 This anthropometric graph shows the distribution of foot size in the general population. How can this sort of data help a company manufacturing work boots decide what sizes to make? (3 marks)

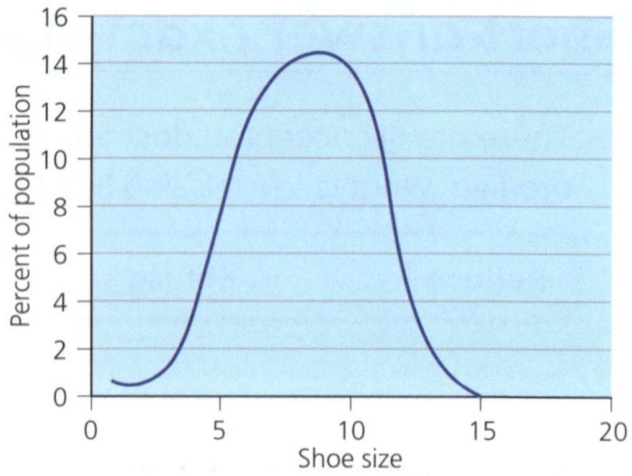

....

....

....

....

....

....

Hint

This question is worth three marks, so make three points.

Sample answer

This graph shows that most people have foot sizes between 6 and 11, so they should make boots in those sizes.

Comment

Do you think that this answer will get all three marks? How could it be improved?

....

....

....

....

6 Choose one of the six Rs and explain what it means for an engineer. (4 marks)

....

....

....

....

....

....

....

....

Hint

Make sure that you think about this in engineering terms, not just in design terms. This will help you get full marks. Giving examples can be a big help in making your answer clear.

Long-answer exam-style practice questions

1 Customer feedback is an important tool in planning future designs for products. Discuss the difference between the feedback you might gain from a final user compared to the feedback from someone involved in the manufacture of the product. (6 marks)

Plan your own answer

Your answer should cover these areas:

- feedback from the final user:
  - how easy or hard the product is to use
  - how it looks
  - how long it lasts
- feedback from production:
  - how easy or hard the product is to make
  - bolts and screws needed
  - use of jigs
  - any fiddly assembly
  - materials.

Sample answer

A company might ask customers for feedback on a product they have bought to help the company make a new version that will sell better. A customer might find it difficult to change the batteries on something and so the designer might change the battery compartment. The manufacturing department could report that the materials used are difficult to get or that the design uses lots of different sizes of screws. These are things the designer could change to make the next version of the product cheaper to make.

Comment

This a fairly good answer but the candidate missed many of the things listed in the plan, for instance the way the product looks. Fashions change and it would be worth the design department revisiting the style of the product on a regular basis.

There won't be any marks for information that is only given in the rough plan, so those possible marks will be lost. It is always a good rule to give some information for each mark for the examiner to tick.

2 A design project could be described as linear or as iterative. Discuss the difference between these two types, and any advantages or disadvantages of them. (5 marks)

....

....

....

....

....

....

....

....

....

....

3 Explain the similarities and differences between quality assurance and quality control. Give examples of both and note how they affect the manufacturing process. (6 marks)

....

....

....

....

....

....

....

....

....

....

....

....

Hint

Always check the number of marks for a question. The examiner is looking to give you marks, so will tick any clear points you make that answer the question: a one-mark question = one tick = one answer. For a six-mark question give five good points – that's five ticks and an extra mark for using clear written language.

Engineering and manufacturing past, present and future

Recall activities

Questions 1–3 are multiple choice questions.

Hint

Simple multiple choice questions can check your broad understanding of a topic. Answer with your best guess, then check the answer. If you don't correct mistakes, you will remember the wrong answer. Checking helps you remember the correct answer.

1 In 1907 the world's first synthetic polymer was invented by Leo Baekeland. What was it called?

- **A** Bakelite ☐
- **B** Biblimorph ☐
- **C** Leolastic ☐
- **D** Nylon ☐

Hint

Examiners often add answers to multiple choice questions that sound possible. Here they have included nylon, which was invented early in the development of polymers, but it isn't the correct answer.

2 Name the electronic device that amplifies electrical signals by controlling the flow of electrons in a vacuum tube.

- **A** Thermistor flow ☐
- **B** Thermionic valve ☐
- **C** Light emitting diode ☐
- **D** Laser maker ☐

3 What do the initials VR stand for in engineering design?

- **A** Visible reality ☐
- **B** Valid registration ☐
- **C** Virtual reality ☐
- **D** Ventral realisation ☐

4 On a separate sheet of paper, draw a timeline for the invention of the computer, from the beginning of the twentieth century to the present day.

Hint

This sort of activity can help you get a feel for the way inventions have been changed and improved over time, with many people adding to the development of a product.

5 The motor car and the washing machine are products that have been manufactured for many years. Choose one and, on a separate sheet of paper, draw a timeline of its development.

Short-answer exam-style practice questions

1 What is the accepted term for the ability of machines to gather information, perform analysis and make autonomous decisions? (1 mark)

....

Hint

This is a one-mark question but make sure you answer with the correct term.

2 Give one use of robotic technology in manufacturing and explain the advantages of this to the company. (3 marks)

....

....

....

....

....

....

....

Hint

It is easy to give an example of a robotic technology (perhaps welding or spray painting) but this will probably get only one mark. There's no need to pick a showy and obscure example, as you won't get extra marks for it. Choose something simple that you have seen in industry, but be sure to explain why the company uses it instead of using skilled operatives.

Sample answer

Spray painting vehicles. The manufacturer will use this because it is quicker than using skilled labour.

Comment

Is this answer worth three marks? Adding another possible advantage (for example, no need for personal protective equipment (PPE), the possibility of 24-hour working or fewer errors) would ensure full marks.

Long-answer exam-style practice questions

Hint

It is a good idea to use a highlighter to make sure you don't miss important parts of the question. In question 1, we have highlighted five key words and phrases. There are five marks available so there will be a mark for discussing each one.

1 The Internet of Things is a name for the way modern technology is **connected** and can **communicate**. Discuss how this can help a company with **several sites** control **manufacturing** and **maintenance**. (5 marks)

Plan your own answer

Your answer should cover these areas:

- connected – using the internet or an intranet
- communicate – sending emails or using shared or cloud-based documents
- several sites – local, national or international
- manufacturing – checking production rates
- maintenance – checking for fault alarms or problems raised by quality control procedures.

....

....

....

....

....

....

....

....

....

....

Sample answer

A large manufacturing company could use connective technology such as the internet or its own internal communication, such as an intranet, to send messages to managers and supervisors who don't even have to be in the same country as the factory. These messages could be about the rate at which the factories are working and whether any problems have occurred. The managers can then send specialist technicians to the sites to sort out anything that has gone wrong.

Comment

This is a very good answer with everything relevant from the plan included. If there is time at the end of an exam, check that all your notes have been included in the answer. In a longer answer, it's better to give more information that is needed rather than less!

2 The introduction of the internal combustion engine revolutionised personal transport with the invention of the motor car. Explain the impact this has had on society, and how current and future developments will affect the design and use of personal transport. (6 marks)

....

....

....

....

....

....

Hint

It's always worth planning out the answer for a longer question. Focus on providing engineering information (for example, use of fossil fuels and replacement technologies), but mention the social impact as well.

In any exam, do not forget to cross out your rough work if you do not want it marked.

3 Describe what is meant by the 'circular economy'. Using examples, from your own experience or research, explain how this impacts on the work of design engineers in industry. (6 marks)

....

....

....

....

....

....

4 The marine engineering sector has been important to the UK economy for centuries. Discuss why this is true. What are the particular challenges faced by marine engineers and designers? (6 marks)

Hint

There are two clear questions here so spend the same amount of time and effort on each one.

....

....

....

....

....

....

....

5 **1** Explain how using virtual reality can augment the industrial design process.

2 Give an example.

3 Discuss the possible advantages and disadvantages. (8 marks)

....

....

....

....

....

....

....

....

....

Hint

This question has three parts, as shown by the **highlighted** numbers. A possible mark scheme might be: **1** = 3 marks, **2** = 2 marks and **3** = 3 marks, so that 3 + 2 + 3 = 8.

Make sure you get every mark!

6 A company is investigating the use of drones to distribute fixings and components around their factory so that technicians do not have to fetch these themselves. What challenges might this represent, and what health and safety issues might arise? (8 marks)

....

....

....

....

....

....

....

....

....

Hint

The examiner will be looking for well thought-out ideas here. You'll have to think carefully about whether an idea is useable or not. Imagine a typical factory environment with doors, high ceilings and overhead wiring when considering your response.

Hint

Watch out for this statement on the front of exam papers:

> Quality of extended responses will be assessed in questions marked with an asterisk (*).

Write these answers in full sentences and use clear vocabulary.

Engineering representations

Recall activities

1 On the diagram identify all the components using their correct names.

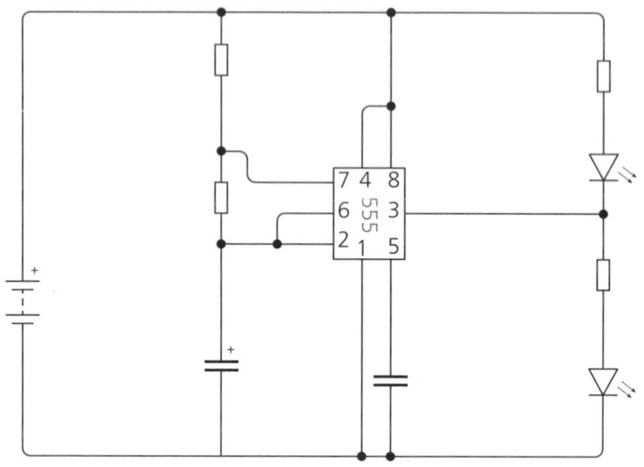

2 Engineering drawings often have coordinate marks round the edge. What is the purpose of these?

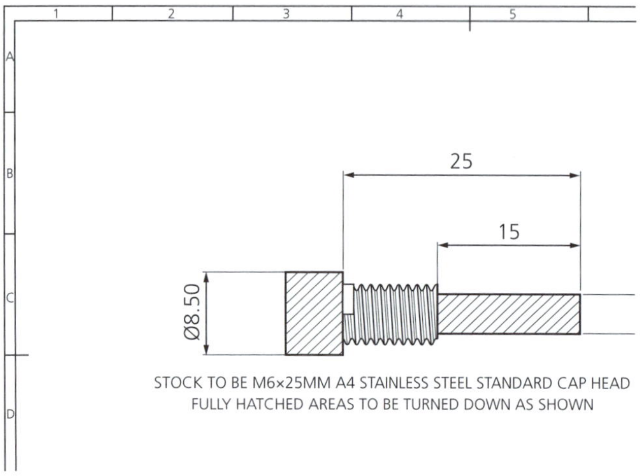

....

....

> **Hint**
>
> You may not know the answer to this right now, which means trying to find out by researching or asking your tutor. This process helps make the information stay in your memory for longer.

3 Find an engineering drawing from your course that uses third angle projection. Use a CAD program of your choice to draw a 3D model of the object. Then produce a third angle drawing of your own. Are there any differences?

....

....

....

Short-answer exam-style practice questions

1 Exploded diagrams are often included in maintenance manuals. Why is this form of drawing useful in this context? (3 marks)

....

....

....

....

2 What does this symbol mean on an engineering drawing? (1 mark)

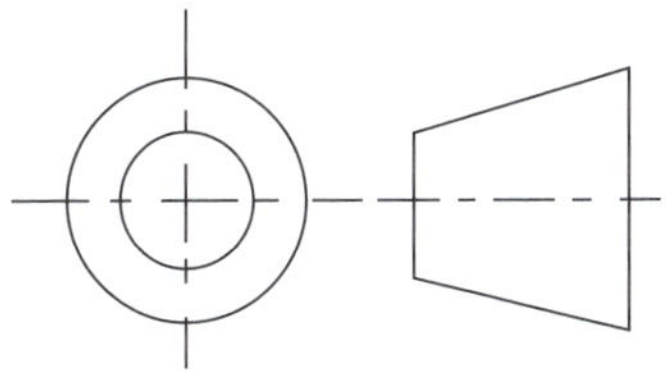

....

....

3 When might a design engineer produce a hand drawn sketch? (1 mark)

....

Hint

Think about the advantages of a quick sketch. When would it be used?

4 CSK is a common abbreviation in mechanical engineering drawing. What does this mean? (1 mark)

....

5 In the space below, produce a system block diagram for the circuit shown here. (2 marks)

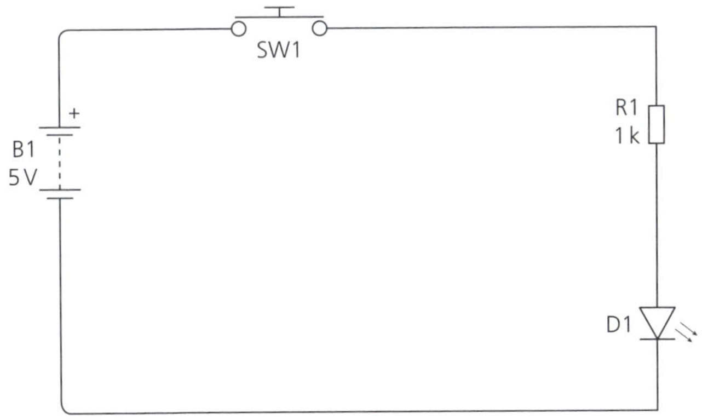

Hint

Block diagrams should be drawn with a ruler and pencil, with arrows connecting the blocks.

6 What does this symbol (right) mean on an engineering drawing? (1 mark)

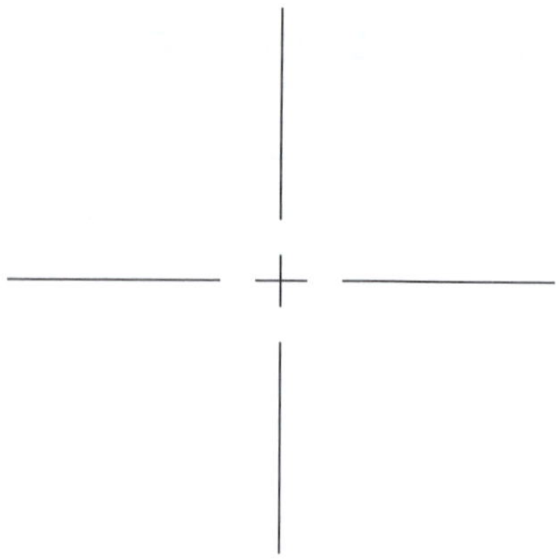

....

7 Calculate the upper and lower limits for the dimension by applying the tolerance shown: 150 ± 0.5 mm. (2 marks)

....

....

Hint

Use your calculator and write the maximum and minimum possible dimensions.

8 A cuboid has been drawn here. What standard drawing technique has been used? (1 mark)

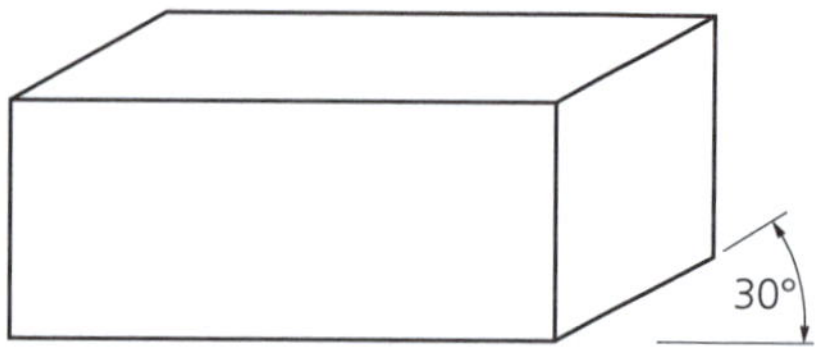

....

9 This drawing has dimensions marked on it. Calculate the diameter of the central circle. (2 marks)

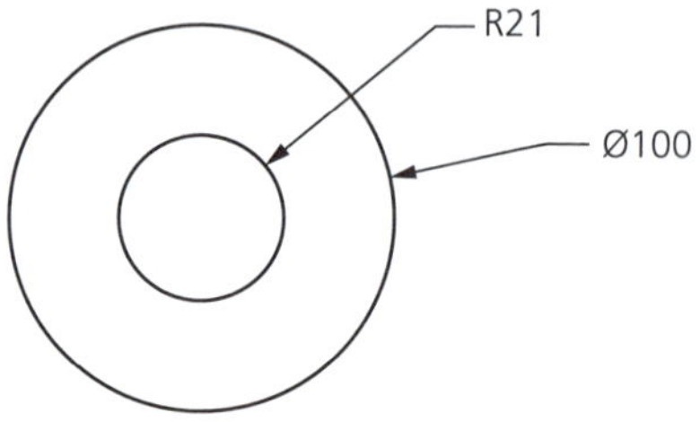

....

....

....

Hint

Be careful with questions which ask you to calculate an answer concerning dimensions. They can look easy, but this one has the two circles dimensioned in different ways. Convert one dimension into the other version to work out the answer.

10 Which of these are valid British drawing standards? Tick two. (2 marks)

- **a** BS 8888 ☐
- **b** BS 9075 ☐
- **c** BS 3065 ☐
- **d** BS 3939 ☐

Hint

Drawing is repetitive and the correct answers have repeats in them.

11 Explain the difference between clearance and interference fits, and give an example of where each would be used. (4 marks)

....

....

....

....

....

....

Sample answer

A clearance fit is loose and gives some room for movement, such as when a pin is used in a hinge. An interference fit is very tight and the components are locked together.

Comment

Is this answer worth four marks? The candidate hasn't given an application for the interference fit (for example, mounting the head on a metal-handled hammer), nor have they explained how the design engineer would indicate the fit required. Taking this feedback into account, write a better answer.

12 Explain why you might use 3D rendered views for a presentation to management personnel, but present isometric and orthographic drawings to engineers. Give an example of another type of drawing that is used for a specialist application. (5 marks)

....

....

....

....

....

....

....

Sample answer

Management staff of an engineering company could be non-engineers. They could be finance people or designers. They would not understand the technical drawings but the fellow engineers would.

Comment

Is this answer worth five marks? The explanation of the first part of the question would gain some marks, but the candidate hasn't given any examples of specialist drawing types. They could have mentioned pneumatic or electronic diagrams, or the structural drawings used by civil engineers.

13 This is an engineering drawing of a turned metal part.

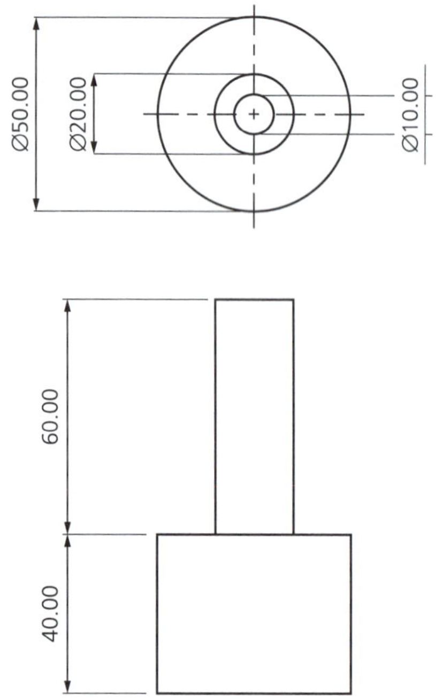

a Draw the hidden detail lines on the drawing. (2 marks)

b Explain why the drawing might include the 3D isometric view of the part. (2 marks)

....

....

....

c If the drawing had information about the dimension tolerances that stated that the length should be as shown ±0.1mm, what is the longest length that would be acceptable? (2 marks)

....

....

d What is the shortest acceptable length? (2 marks)

....

....

14 This is the circuit diagram of a power supply circuit.

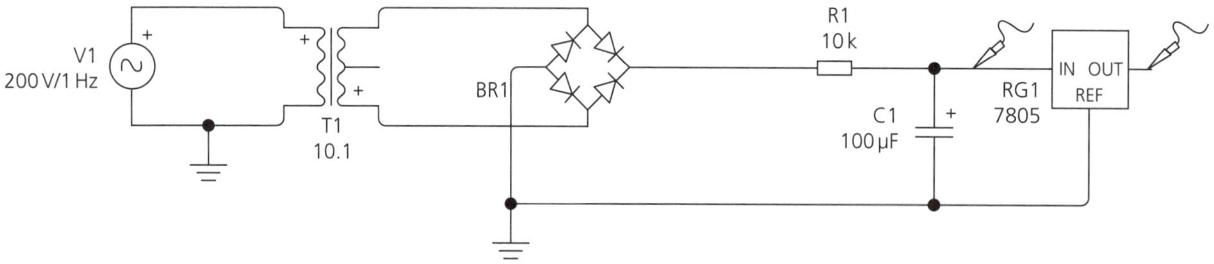

Explain why it is necessary to use standard electronic symbols on a circuit diagram like this. (6 marks)

....

....

....

....

....

....

....

Long-answer exam-style practice questions

1 The use of computer aided design has increased over the past few decades. Discuss why this might be so, and explain the advantages and disadvantages of CAD to an engineering company. Write your answer on a separate sheet. (8 marks)

Plan your own answer

Your answer should cover these areas:

- increasing use of CAD:
  - most work is on computer
  - files can be easily stored and shared
  - computer hardware has improved steadily
- advantages of CAD
  - accurate
  - produces clear drawings
- disadvantages of CAD
  - requires use of a computer
  - can only be used by trained individuals.

The plan has seven points, so should gain a tick for each point. Now string these points together into a well written answer for the final mark.

Hint

Eight marks is a lot for a seemingly simple question. Plan your answer to make sure you give seven clear points and that the answer is well written to gain full marks.

Sample answer

As the standard of computer hardware has improved over the last few decades, it has been possible for computer aided design to improve as well. Files created on CAD can be saved for security of the data and to enable the drawings to be shared in collaborative working. The other advantages of CAD are the accuracy and quality of the drawings, with errors being easily changed and older versions kept for reference. The only disadvantages to the company would be the need for a good enough computer and screen to work with the software, and the fact that the users need to be trained on the software and kept up to date with new versions as they come out.

Comment

This is a very good answer. The candidate has missed out point 1 from the plan but would gain two marks from point 2. The rest of the answer is excellent with some good detail added. Go through the answer yourself and tick the points as you find them.

Write your own answer using this plan.

2 A customer has asked for a new steel-machined part is to be manufactured and it is the job of a production engineer to check that the part will be made to the tolerances given. Explain why this is important to both the customer and the company itself. Write your answer on a separate sheet. (6 marks)

Essential mathematics for engineering and manufacturing

Hint

You might not like maths, but it is very important. Never be tempted to say 'I'll never use maths at work!' Maths is involved in most projects in engineering.

There are several maths topics you need to know for Unit 4. The questions will usually be asked in an engineering context, so read the questions carefully, spot the maths you need to use and then work out the answer. Don't forget to show your workings!

Recall activities

The best way to get your maths brain working is to try some quick questions.

1 **a** Write out what BIDMAS stands for.

....

b Using BIDMAS work out the following maths questions. Try not to use your calculator for the first few.

$5 \times 5 + 3 \div 2 =$ ....

$5.5 + 2.5 + 4 \div 2 =$ ....

$(6 \times 3) + (8 \times 4) =$ ....

$(210 \div 3) - (5 \times 6) =$ ....

2 Fill in the following numbers to fit the description at the top of the table. The first one has been done for you. Round up or down when necessary.

Number	4 significant figures	3 decimal places
498.29426	498.3	498.294
2.583579		
1086723.44		
2347.9599		
0.238723		

3 Show 25% as a fraction.

....

4 Rearrange this equation to simplify it.

$s = 12d + 22d - (35 - 13)d$

....

5 Rearrange this equation to show how to work out y.

$x = 5y - 6e + 8$

....

6 Solve these simultaneous equations.

$2x + y = 5$; $3x + y = 7$

....

....

7 Factorise and solve this quadratic equation.

$2x^2 - 6x - 8 = 0$

....

....

8 Calculate:

a $5 \times 5^2 \times 5^0 =$ ....

b $4^{-2} =$ ....

9 Simplify:

$f^2 \times f^2 =$ ....

10 Calculate:

a $\log(10^2) =$ ....

b $\log(10^{-2}) =$ ....

Simplify:

c $\log A - \log B =$ ....

d $\log 10^6 + \log 10^3 =$ ....

e $\ln(x + 2) - \ln(x + 1) =$ ....

11 What is the next term in these sequences?

a 1, 3, 5, 7, 9, ...

b 2, 3, 5, 7, ...

What is the next term in this arithmetic progression?

c 1, 4, 7, 10, 13, ...

d Find the 6th term of the following series: 6, 12, 18, 24.

....

What is the next term in these geometric progressions?

e 1, 2, 4, 8, 16, ...

f 2, 6, 18, 54, ...

g Find the 9th term of the following series: 5, 10, 20, 40.

....

h Find the nth term of the following series: 3, 15, 75, 375, 1875.

....

Short-answer exam-style practice questions

1 Solve these expressions.

a $5 \times 5 + (4 \div 2) =$ .... (1 mark)

b $3^2 \times 4 + 6 \div 3 =$ .... (1 mark)

c $(5.5 + 2.5 + 4) \div 2 =$ .... (1 mark)

d $2^2 \times 3^2 \times 5 - 11 =$ .... (1 mark)

e $(4^2 + 2) \times 3 \div 2 + 2 =$ .... (1 mark)

2 A drawing shows a dimension at a 10:1 ratio. The dimension is 23 mm. What is the measured length of the line on the drawing? (1 mark)

....

3 What is $\frac{3}{4}$ as a percentage? (1 mark)

....

4 What is 30% + 60%? (1 mark)

....

5 **a** Rearrange and simplify $k = 6d + 10 - 2d + (22 + 4)d$ (1 mark)

....

b Expand $(x - 5)(x + 8)$ (1 mark)

....

c Factorise $x^2 + 10x + 25$ (1 mark)

....

6 Rearrange the following equations to work out y.

a $x = y + 2$ (1 mark)

....

b $x = 5 + y$ (1 mark)

....

7 Solve the simultaneous equations. (2 marks)

$2x + 3y = 8$; $3x - y = 23$

....

....

8 Expand the brackets. (1 mark)

$(x + 2)(x + 3)$

....

9 Factorise the expression. (2 marks)

$x^2 + 4x - 12$

....

10 Factorise and solve the following quadratic equations.

a $3x^2 + x - 10 = 0$ (2 marks)

....

....

b $4x^2 - 13 = 3$ (2 marks)

....

....

11 Solve the following expressions.

a $3^3 + 6^2 =$ .... (1 mark)

b $(9)^{\frac{1}{2}} =$ .... (1 mark)

12 A car tyre has a diameter of 40 cm. The car moves forward for 25 m. How many times has the wheel gone round? (2 marks)

....

....

13 Simplify the following expressions.

a $g^5 \div g^2 =$ .... (1 mark)

b $(k^2)^3 =$ .... (1 mark)

14 Calculate:

a $\log(10^{4.3}) =$ .... (1 mark)

b ln(6) = ... (1 mark)

c ln(4.2) = ... (1 mark)

15 a This shape in the diagram is the cross section of an aluminium extrusion. Calculate the area of the cross section. Show your workings and give your answer in millimetres squared (mm^2). (3 marks)

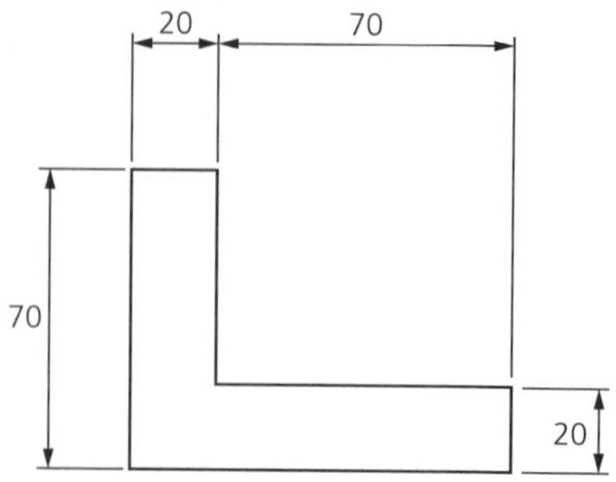

....

....

....

Hint

As the question tells you explicitly what unit to use, you should definitely use it. Three marks means three things to get right, so show your working. Calculate the area of the shape by chopping it up into two rectangles and adding to get the total area, or (more unusually) by calculating the area of the outside rectangle and then subtracting the inner.

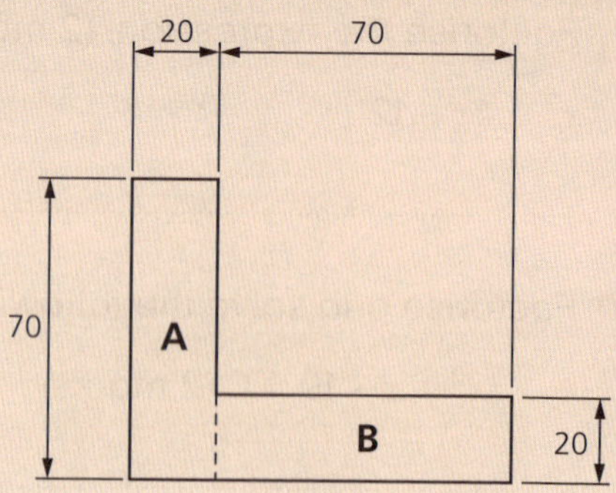

area A = $20 \times 70\,mm^2$

area B = $20 \times 70\,mm^2$

total = $1400 + 1400 = 2800\,mm^2$

b The shape in part a is extruded to 100 mm long. Calculate the volume of the complete extrusion. Express your answers in mm^3. (3 marks)

....

....

16 Solve this quadratic equation for x. (2 marks)

$4x^2 - 5x - 12 = 0$

....

....

17 Calculate the area of the shape in the diagram. Express your answer in mm^2. (3 marks)

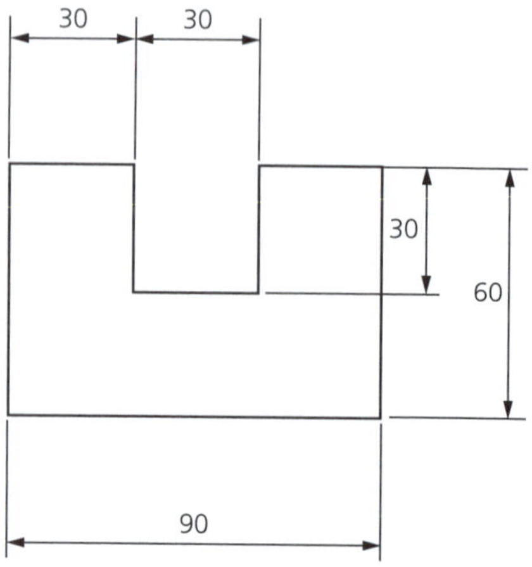

....

....

....

....

18 Calculate the area of the shape in the diagram. Express your answer in mm^2. (3 marks)

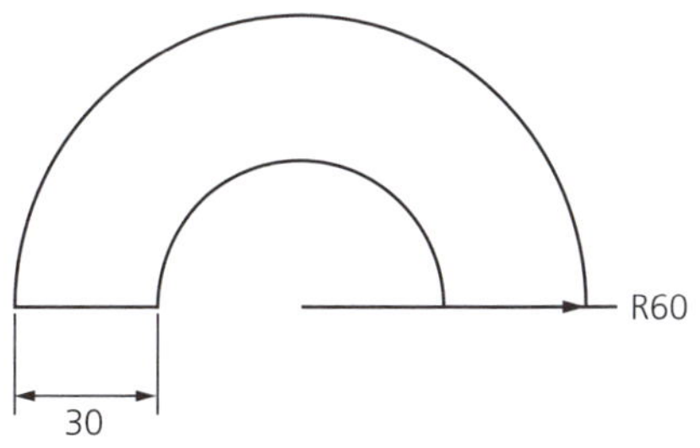

....

....

....

19 A small engineering company has been asked to make a batch of 100 steel plugs.

Each plug is 200 mm long and has a 50 mm radius. Calculate the volume of material needed to make 100 items. (3 marks)

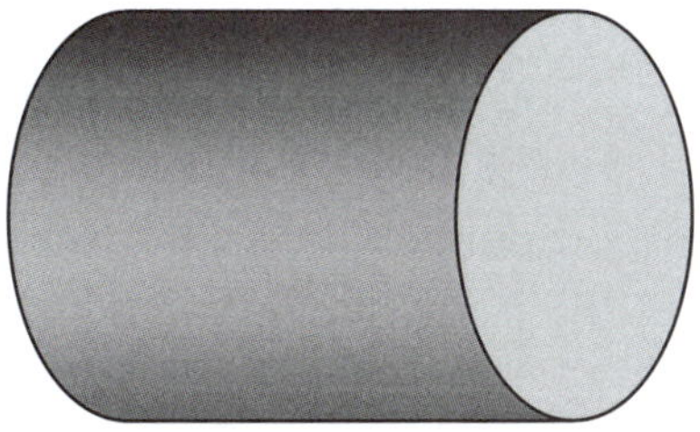

....

....

....

20 The cone in the diagram has to be cast in concrete for an art installation.

The cone is 1.6 m tall and has 200 mm radius at the base. What volume of concrete will be needed? (3 marks)

....

....

....

21 This graph plots the length of a bolt (y-axis) against the thread diameter (x-axis).

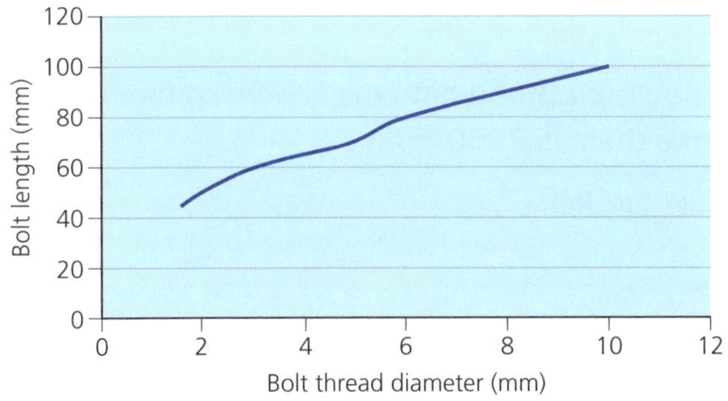

Hint

There are three marks for this question, but two correct answers would only give two marks. The other mark must be for using the correct units. Note that this question also checks that you understand what the 'M' in M12 means (12 mm thread).

Estimate the length of an M8 bolt. Now estimate the length of an M12 bolt. Both answers should be given in whole millimetres (mm). (3 marks)

....

....

....

22 This graph plots the coordinates (–2, –3), (–1, –1), (0, 1), (1, 3), (2, 5) and (3, 7).

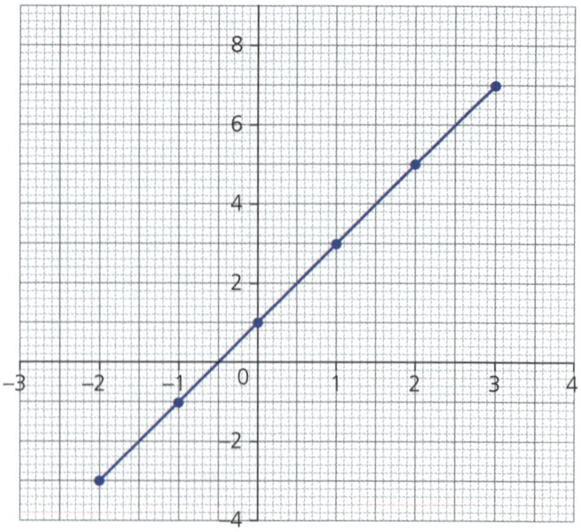

Work out the values for m and c in the formula $y = mx + c$ for this graph. (2 marks)

$m =$ ....

$c =$..................................

23 In a tinning factory there is a full hopper containing 2500 l of tomato sauce. During a large production run of baked beans, 1250 l of sauce is used. The production machinery leaks 50 l of the sauce on to the floor. Show:

a the fraction of the original full hopper used in the production run

b the fraction of the amount used for the production run that is lost in the leak

c the amount wasted from the original full hopper as a percentage. (4 marks)

....

....

....

....

Hint

Read the question carefully. Part a asks for a fraction of the original amount, but part b requires a fraction of the quantity used in production. Part c is referring to the original amount and asks for a percentage.

24 A manufacturer of wheel bearings needs to count how many ball bearings they have in stock. Counting them one at a time would be time consuming and could be prone to error. They decide to weigh them instead, to the nearest 0.1 g.

A box of 6 mm ball bearings weighs 0.9106 kg. They are poured onto the scales so that the box is not weighed. The density of the steel used to make them is 8050 kg/m^3.

Estimate how many ball bearings are in the box. (4 marks)

....

....

....

....

....

Hint

Do this question one step at a time. Do not round up or down until you get to the final answer. This final number has a unit, so now you can apply normal rounding up or down procedure. A question like this tests how you cope with very small and very large numbers.

Sample answer

volume of one ball bearing $= \frac{4}{3} \times \pi \times r^3$

$= \frac{4}{3} \times 3.142 \times 3 \times 3 \times 3 = 113112$

volume of the one ball bearing in $m^3 = 0.000113112$

weight of the ball bearing in kg = density × volume

$= 8050 \times 0.000113112$

$= 0.0009105516$

number of ball bearings $= \frac{0.9106}{0.0009105516} = 1000.053155$

Estimate that there are 1000 ball bearings in the box.

Comment

Would you give this answer full marks? Adding the units to all stages would make it clearer.

volume of one ball bearing $= \frac{4}{3} \times \pi \times r^3$

$= \frac{4}{3} \times 3.142 \times 3 \times 3 \times 3 = 113112\,\text{mm}^3$

volume of the one ball bearing in $m^3 = 0.000113112\,\text{m}^3$

weight of the ball bearing in kg = density × volume

$= 8050 \times 0.000113112\,\text{kg}$

$= 0.0009105516\,\text{kg}$

number of ball bearings $= \frac{0.9106}{0.0009105516} = 1000.053155$

Estimate that there are 1000 ball bearings in the box.

25 A new welding robot was installed to weld the frames of bicycles together. The graph shows the number of frames welded against the number of weeks the robot has been working.

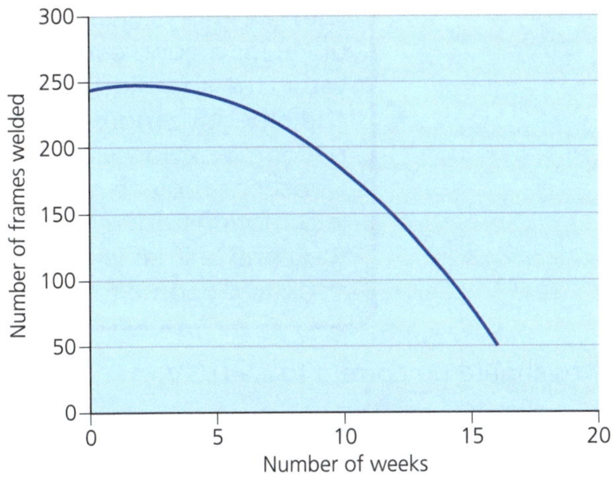

Unfortunately the data that was used to plot this graph has been lost and we only have the formula for the curve left to use as evidence to the supplier. You must work out using differentiation the maximum count the robot achieved and in which week that occurred.

The formula is $y = 4x - x^2 + 243$. (4 marks)

....

....

....

26 You have been asked to make a small batch of the brackets shown in the diagram for testing before they go into production.

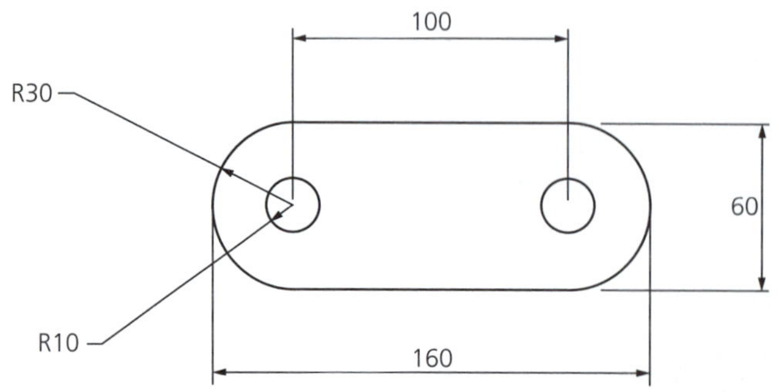

> **Hint**
>
> Do you think the two radii are concentric? Is there anything to back that up on the drawing?

Interpret the drawing to calculate:

a the diameter of the drill required to make the holes

b the area of the bracket as shown

c the volume of the metal used in the bracket if it was 5 mm thick

d the area of a new bracket 100 mm longer than the drawing. (4 marks)

....

....

....

....

27 A capacitor is charging in a circuit and the technician measures the voltage every second. This is the resulting graph.

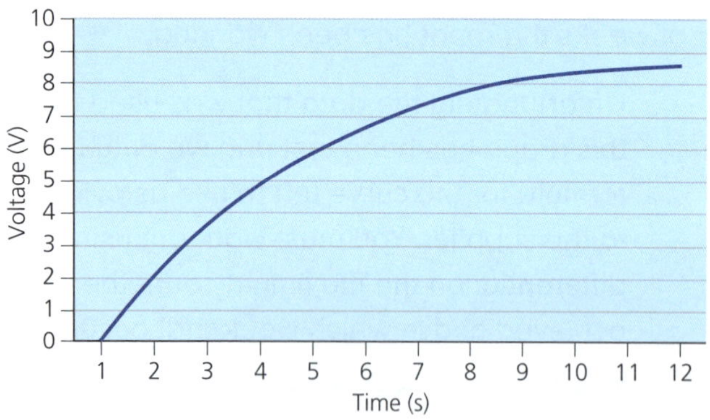

> **Hint**
>
> The four marks for this question are probably split with two marks (including units) for a good estimate within the specified range. The other two marks will be for the explanation. Questions may cover information from more than one unit, so you might need to do some research.

Estimate the value of the voltage at 6 s. Your answer should be correct to +/–0.3 V.

Use your knowledge of electronics to explain why the voltage might never go above 9 V. (4 marks)

....

....

....

....

28 A company specialising in restoring historic aeroplanes has to fit a seat into the cockpit of an aircraft.

The bolt on the seat is 5/16" wide. Unfortunately, the technician has only been supplied with a metric drill set as shown in the table.

Hint

Obviously, the best solution would be to purchase an Imperial drill set with all the right sizes but this is not always possible.

Drill sizes available (mm)													
2	3	3.2	4	5	5.5	6	6.5	7	7.5	8	10	12	16

a Using the conversion factor of 1"(inch) to 25.4 mm, suggest the nearest metric drill size for the bolt to pass through and be a reasonable clearance fit. (2 marks)

....

....

b Another hole must be drilled for a bolt of size 5/8". Suggest a good metric drill size for this hole. (2 marks)

....

....

c Wiring needs to pass through from behind a bulkhead (wall). The cable is 1/4" in diameter. Which drill would be best for this application? (2 marks)

....

....

Sample answer

The sample answer shows the workings for part a.

$$5/16 = 0.3125 \text{ inches}$$

$$\text{convert to mm} = 0.3125 \times 25.4$$

$$= 7.9375 \text{ mm}$$

A 7 mm drill hole would be too small but an 8 mm would allow the bolt to go through with a good clearance.

Correct drill size = 8 mm.

Comment

Now try parts b and c. There are two marks for each of these parts in addition to the two marks for part a.

29 An electronics engineer has a circuit similar to the one shown in the diagram.

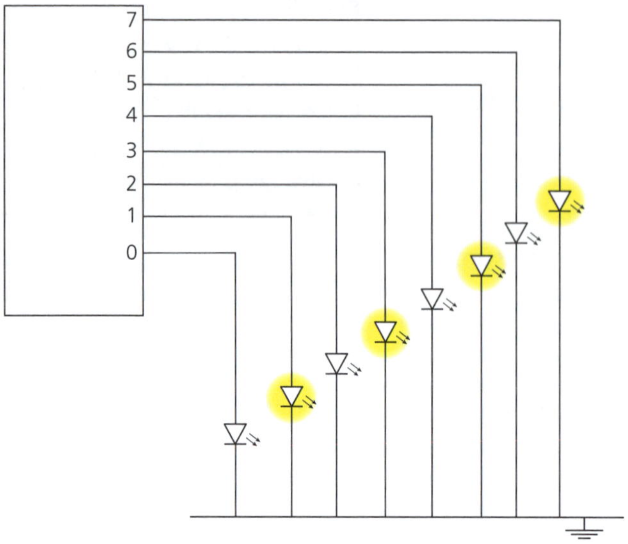

An integrated circuit controls eight LEDs by setting the numbered pins (0 to 7) to either a one (3 V) or a zero (0 V).

They display a pattern where alternating LEDs are lit. The LED at 0 is not lit in this pattern to start with.

a Write this number as a binary, a decimal and a hexadecimal number. (2 marks)

....

....

....

b Now the engineer wants to change the display so that the LEDs are lit in a new arrangement as shown in the next diagram.

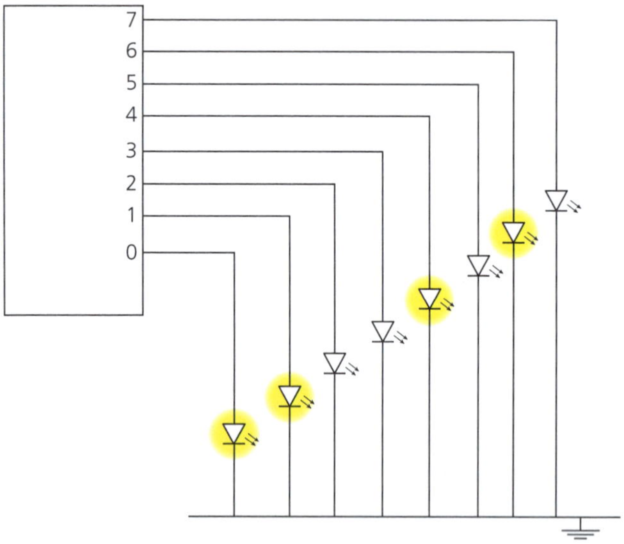

Write this new number as a binary, a decimal and a hexadecimal number. (2 marks)

....

....

....

Sample answer

10101010

AA

170

Comment

For part a, write down the 0s and 1s in the correct order with 0 as the smallest significant figure and 7 as the largest, so the number is 10101010. In hexadecimal that is AA. In decimal it is 170.

Now work out part b.

30 Data analysis is a valuable tool for a data control department.

The quality control department at a manufacturer has been given the set of data in the table to analyse. They need to know the mean, mode, median and standard deviation of these numbers. Work these out and complete the report below. Plot the data on the graph provided. A space has been left for your calculations. (8 marks)

Count	1	2	3	4	5	6	7	8	9	10
Total	3	7	12	22	31	33	17	6	4	1

Mean = ...

Median = ...

Mode = ...

Standard deviation = ...

Hint

Look at the number of marks available. The mean, median and mode would seem to be an easy three marks so, how do the other marks split? Working out standard deviation is quite long winded so two marks for that, leaving three marks for the graph? A detailed breakdown of marks is not provided, so you have to make an educated guess here.

31 Aluminium is available in 61 cm × 122 cm sheets. Calculate how many 300 mm × 300 mm squares can be cut from a sheet of these dimensions. (2 marks)

....

....

32 The diagram shows the movement of a lever which is part of a robot arm.

The distance from point A to the pivot is 10 mm.

The distance from point B to the pivot is 40 mm.

If point A moves 10 mm to the right, how far would point B move to the left? (2 marks)

....

....

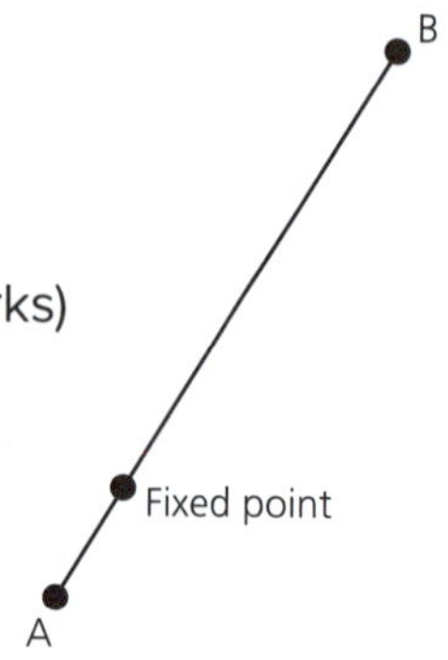

33 This is a list of all the weights of cakes produced on a production line over five minutes of operation. The cakes are weighed in grams.

120	121	119	120	122	123	121	118	119	121

a Calculate the mean, median and mode of this sample. (2 marks)

....

....

b Now calculate the standard deviation of this sample. (2 marks)

....

....

34 An installer wants to work out the maximum number of solar panels that can be fitted on a domestic garage roof. Each panel is 1 m × 2 m and the roof is 10 m long by 3.5 m high. What is the maximum number of panels that will fit? (3 marks)

....

....

....

35 A small building needs painting on all four walls (but not over the roof).

If the building is higher at the back than the front, as shown in the diagram, calculate the area of the four walls.

If 1 l of paint covers 5 m^2 how many full tins of paint will be needed? (4 marks)

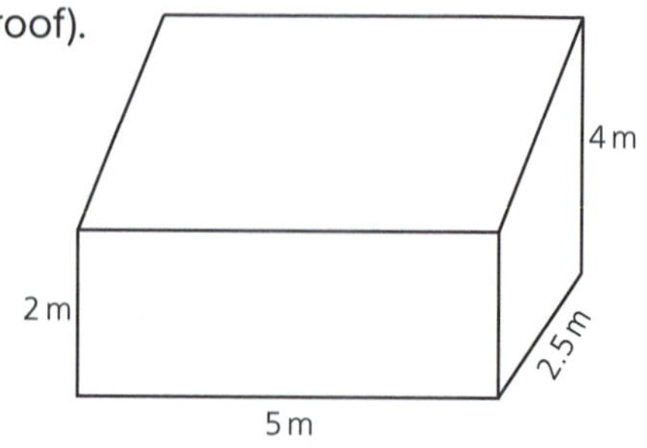

....

....

....

Essential science for engineering and manufacturing

Hint

Engineering is often described as 'putting science into action'. Knowing the scientific principles underpinning any engineering project can help you to understand *how* and *why* a process works. This knowledge comes not just from physics, but also from chemistry and biology.

Recall activities

1 Match each of the six SI base units to the quantity being measured.

Unit of measurement
Mole (mol)
Seconds (s)
Metre (m)
Candela (cd)
Ampere (A)
Kelvin (K)

Quantity being measured
Luminous intensity
Electric current
Thermodynamic temperature
Length
Amount of any substance
Time

2 Convert the following Imperial measurements into standard metric units.

a 2 yards = ...

b 3lb 2oz = ...

c $\frac{1}{2}$ gallon = ...

3 How many zeroes are there after the 1 when 1 GW (gigawatt) is expressed in watts?

....

4 How many zeroes are there after the decimal point when 1 pV (picovolt) is expressed in volts?

....

5 Which of these quantities are vectors and which are scalars? Tick the relevant box.

Quantity	mass	velocity	gravity	distance	energy	weight	power
Scalar							
Vector							

Quantity	force	momentum	time	acceleration	displacement	volume
Scalar						
Vector						

6 Using a copy of the Periodic Table, write down the symbols for the following elements to find a word that is the 'enemy of revision'.

Praseodymium Oxygen Chromium Arsenic
Titanium Sodium Titanium Oxygen Nitrogen

....

7 Solve the key terms crossword. This is not a complete list of the key terms you should be familiar with, so check the textbook for any that are not included here.

ACROSS

1 Two or more atoms bonded together
5 The measure of a system's thermal energy
6 Regions of a fluid where the flow revolves
7 Amount of three-dimensional space occupied
10 Deposit a metal on a part
11 Finding the precise value of a variable
12 Transfer of heat via electromagnetic waves
13 Mass × velocity of an object
14 Transfer of heat by direct contact
15 Downwards force that opposes lift
16 An upward force exerted by a fluid
17 A state where forces balance

DOWN

1 Measure of an object's inertia
2 Negative terminal of an electrolytic cell
3 Force applied perpendicular to the surface
4 Force that makes an object move forward
8 Negatively charged subatomic particles
9 Margin of doubt about a measurement
10 Process that causes chemical decomposition

8 You are asked to test the run out of an automotive brake disc mounted on a lathe. The measurement is expected to be between 0.1 mm and 0.2 mm. Which of the following would be the best measuring tool to choose?

A Dial gauge ☐

B Steel rule ☐

C Digital thermometer ☐

D Vernier calliper gauge ☐

> **Hint**
>
> Actually *using* the tools mentioned in your textbook is the best way to remember which one is which. Always use the correct terms when working in the workshop.

9 Arrange these metallic elements in order of reactivity, from most reactive to least.

Magnesium	**Copper**	**Lithium**	**Gold**	**Tin**	**Iron**

....

10 Decide what sort of motion defines each of these activities. Some might have two different types of motion, so tick two if applicable.

Activity	Linear	Rotary	Reciprocating	Oscillating
Pendulum of a clock				
Frisbee flying through the air				
The piston of an internal combustion engine				
Hands on an analogue clock				
A child on a swing				
The end of the saw when sawing				
Train travelling down a track				

11 A soldering iron is switched on. Describe how the heat from the iron is transferred to the solder on a circuit. Use the correct term or terms from the box.

Conduction	**Convection**	**Radiation**

....

....

....

12 Choose another way heat can be transferred and describe a situation where that might happen.

....

....

....

13 2.00 l of a gas is at 740.0 mm Hg pressure. Using Boyles's law calculate its volume at 760 mm Hg pressure.

....

....

....

14 A balloon is fitted over the top of an empty bottle as shown in the diagram. When the bottle is held in a bowl of hot water the balloon inflates. Why is this?

....

....

....

Short-answer exam-style practice questions

1 Describe the differences between a primary and secondary chemical cell. (3 marks)

....

....

....

2 Choose one application for electroplating and describe the process and its advantages. (3 marks)

....

....

....

....

3 A sack of injection moulding granules weighs 9 stone. Staff need to know the weight in kilograms. Convert the weight into kilograms. (3 marks)

....

....

....

4 A resistor in a circuit produces a voltage drop of 6 mV when 10 μA is passed through it. What is the value of the resistor in ohms (Ω)? (2 marks)

....

....

5 A professional quality steel rule does not usually have any blank space in front of the zero point on the scale, as shown in the diagram. Why is this? (2 marks)

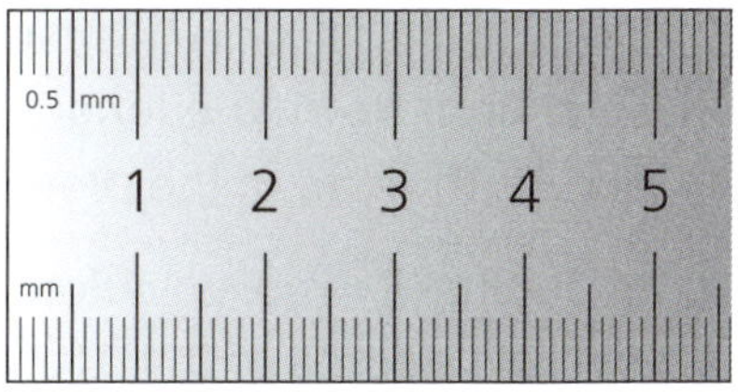

....

....

....

6 A Vernier calliper gauge is used for making many measurements in an engineering setting. Describe one application and explain why the Vernier is suitable for this job. (3 marks)

....

....

....

7 You have been asked to check the length, height and diameter of the through hole on this suspension bush to 0.1mm.

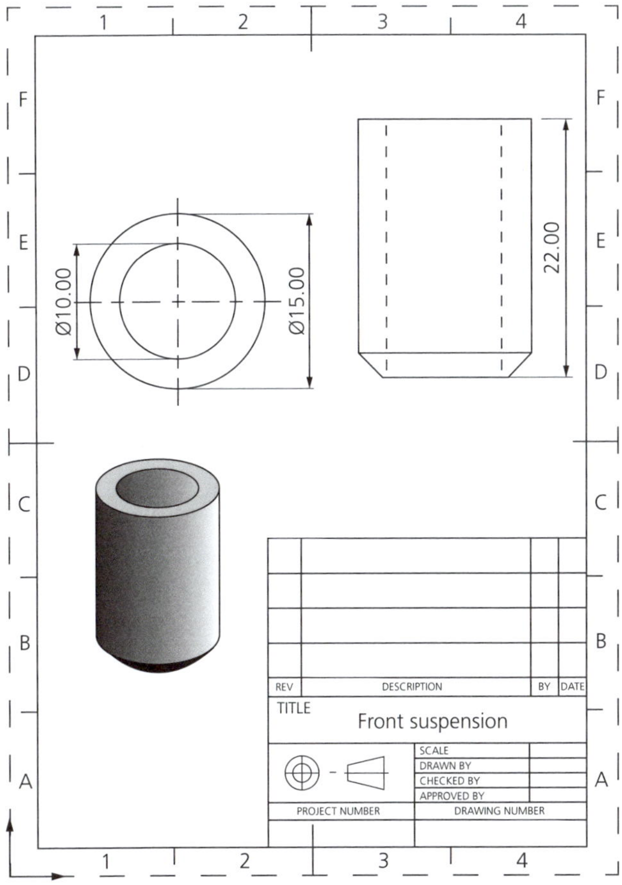

What piece of measuring equipment would you choose and why? (3 marks)

....

....

....

8 Draw the air flow in this pipe. Assume the flow is left to right, is laminar down the length and then becomes turbulent as it leaves the end of the pipe. (4 marks)

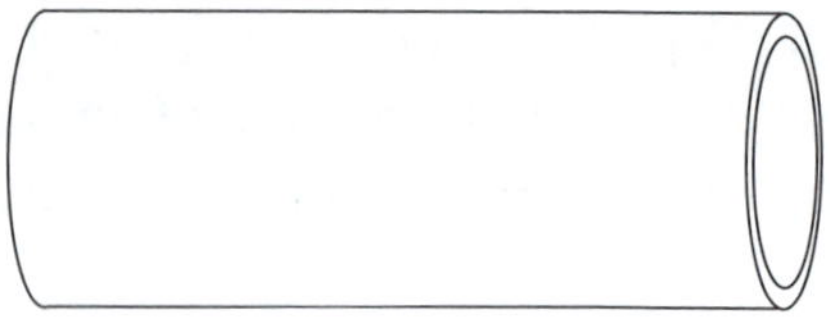

9 **a** Using $F = ma$, calculate the acceleration of a mass of 10 kg with a force of 3 N acting on it. (2 marks)

....

....

b Using $s = ut + \frac{1}{2}at^2$ and knowing the mass is stationary at the start ($u = 0$), how far will the mass travel in 3 seconds? Remember, t = time and a = acceleration. (2 marks)

....

....

10 Two hydraulic cylinders are attached by a length of tube. If the larger one has a diameter of 30 mm and moves 20 mm, how far will the other cylinder with a diameter of 10 mm move? (4 marks)

....

....

....

....

11 A force of 600 N is applied to an area of 1000 mm^2. Calculate the pressure. (3 marks)

....

....

....

12 Explain how a fuel tank might be protected from corrosion using galvanic protection. (4 marks)

....

....

....

....

13 Explain why using a lubricant in a mechanical system can help overcome friction. (4 marks)

....

....

....

....

14 Match the key terms to the definitions. (6 marks)

Term
Anode
Cation
Dissociation
Anion
Cathode
Electrolysis

Definition
Positively charged ion
Negatively charged ion
Process that causes chemical decomposition through the use of electricity
Breaking up a compound Into simpler components
Positive electrode
Negative electrode

15 Heat can be transferred from one body to another by radiation. How is the heat transferred during this process? (2 marks)

....

....

16 Describe the difference between sensible and latent heat. (4 marks)

....

....

....

....

17 Complete this short paragraph about the Kelvin (unit K) and Celsius (unit °C) temperature scales by filling in the missing words and numbers. (3 marks)

A body is said to be at .... zero when its temperature is 0 K. This temperature is equal to –277.15 °C. If a pan is on a stove and is being kept at 60 °C, its temperature is .... This temperature is known as the .... temperature of the pan.

18 **a** A gas bottle containing butane gas is labelled as being at 170 kPa. Atmospheric pressure is 98 kPa. What is the absolute pressure of the gas in the bottle? (2 marks)

....

....

b When the gas is let out of the bottle, the temperature of the bottle goes down. Explain why this happens. (3 marks)

....

....

....

19 A person walks 100 m east, turns left and walks 50 m north. The person then turns left again and walks west for another 30 m.

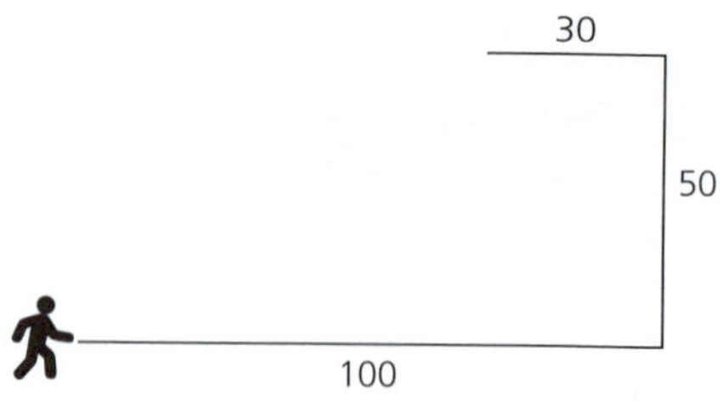

a What distance has the person walked? (2 marks)

....

....

b What is their displacement at the end of their walk? (4 marks)

....

....

....

....

Hint

You will need to work this out using trigonometry.

Long-answer exam-style practice questions

1 Scientists are researching the possible use of hydrogen as a replacement for fossil fuels in automobiles. As this development gathers momentum, there will be a need for infrastructure to manufacture, store, transport and sell large quantities of hydrogen gas. Discuss the advantages of hydrogen as a fuel, any challenges or problems with its use and any possible solutions. (8 marks*)

Plan your own answer

Your answer should cover these areas:

- advantages – abundant, can be manufactured
- problems with hydrogen – flammable, very small molecules that escape easily
- possible solutions

....

....

....

....

....

....

....

....

....

....

Hint

Use this protocol to make sure you answer a longer question fully and correctly:

- **Highlight** key words in the question
- **Plan** your answer
- **Check** your plan against the question
- **Write** your answer
- **Check** your answer against the question.

Now write your own answer including the details missing from the sample.

Sample answer

Hydrogen is the most abundant element in the whole Universe. It can be used to drive internal combustion engines with small changes to the engines and this makes it a good alternative to fossil fuels.

Hydrogen can be made from many different sources and can be made using energy from renewable sources, such as solar or wind power. The main problem with hydrogen is that it is a very small molecule and can leak from storage tanks and is hard to seal. Tanks have to be specially made and will be very expensive.

Comment

The details included in the answer are good, but it doesn't address the whole question. It is easy to write a plan and then follow it without rechecking the question. Here the candidate has made some good points but missed out any possible solutions to the problems or challenges with its use.

2 A loose material can be compressed into shapes suitable for packing fragile items. One cubic metre weighs 2.2 kg.

a Calculate the density of the loose material. (2 marks)

....

....

b The material is compressed into one-twentieth $\left(\frac{1}{20}\right)$ of the original volume using a press. What is its density in the compressed state? (2 marks)

....

....

c How much would one cubic metre of this compressed material weigh? (2 marks)

....

....

3 Explain the difference between the Cartesian and the polar coordinate system. You may use diagrams in your explanation. (6 marks)

....

....

....

....

....

....

....

Hint

Draw diagrams using a pencil and ruler. Draw lightly at first in case you need to rub out, then go over the lines when you are confident.

4 A student is conducting experiments with helium-filled balloons. They have some questions to research.

a Why does a balloon filled with helium rise in the atmosphere if it is released? (2 marks)

....

....

b What will happen to the balloon as it rises into the upper atmosphere? (2 marks)

....

....

c Explain how a balloon filled with helium moves in an accelerating car. (2 marks)

....

....

Materials and their properties

Hint

A material is anything that you can use to make something else. Items from silk cloth to steel girders are all made of materials.

Recall activities

1 Write out one key difference between physical and mechanical properties.

....

2 Match each mechanical property to its description.

Mechanical property
Toughness
Hardness
Plasticity
Brittleness
Elasticity
Ductility
Malleability

Description
How well a material resists surface indentation, scratching or abrasion
The amount of impact energy a material can absorb until it fractures
Qualitative description of low toughness in materials that are not impact resistant and shatter easily
The ability of a material to undergo plastic deformation without failure when subjected to a tensile stress
Qualitative description of the ability of a deformed material under load to recover its original shape
Qualitative description of the ability of a deformed material under load to retain that deformation permanently
The ability of a material to undergo permanent plastic deformation when subject to a compressive stress without failure

3 What is the Wohler test testing for?

F..

4 Which mechanical property does the Rockwell test for?

H..

5 Why would a piece of iron railing need a protective coating?

....

6 Which of these are thermoplastic polymers and which are thermosetting? Tick the correct column.

Polymer	Thermoplastic	Thermosetting
Urea formaldehyde		
Polypropylene (PP)		
Phenol		
Polylactic acid (PLA)		
Epoxy resin		
Polyester resin		
Polycarbonate		
High impact polystyrene (HIPS)		

7 The properties of iron change depending on what metal is in the alloy and the carbon content of the mixture. If the iron alloy has significant other content, it is called steel.

For each of these alloys, find out the major properties and give one use for each type.

Type of steel	Properties	One use
High-carbon steel		
Stainless steel		
Cast iron		
Low-carbon steel		
Medium-carbon steel		

8 Draw a cubic crystalline structure, such as might be found in a metal sample. Why does this structure give the metal strength?

....

9 What is the name of the bond that is found in a compound of one carbon atom and four hydrogen atoms? Draw the structure and state the name of this compound.

....

10 Tick to show whether each of these materials is an element or a compound.

Material	Element	Compound
Copper		
Acrylonitrile butadiene styrene (ABS)		
Carbon		
Medium density fibreboard (MDF)		
Glass		

11 State the difference between a softwood and a hardwood, and explain how this difference is produced by the way the trees grow.

....

....

12 Explain the difference between brazing and welding.

....

....

13 A circular, solid cross-section suspension strut on a motorbike has a diameter of 23 mm and supports an axial load of 1000 N. Calculate the stress on the strut.

....

....

Short-answer exam-style practice questions

1 A cube of a thermosetting polymer has sides 50 mm in length. It weighs 160 g. Calculate the density of the polymer in g/cm^3. (2 marks)

....

....

....

Hint

Danger! The use of the word 'thermosetting' is included to put you off. It doesn't matter what the material is, this is a simple density calculation. The bigger problem here is to convert the units.

2 Sketch how a piece of steel under tension might look at the point in the graph between E and F, the 'necking zone'. (2 marks)

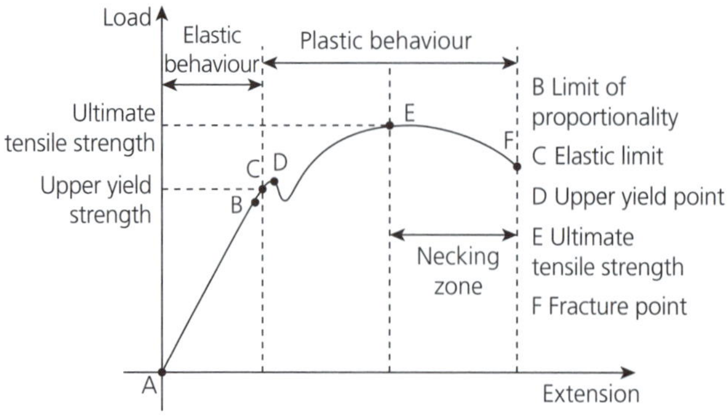

Hint

In your textbook, look up a diagram showing Hooke's law and copy it carefully. Use a pencil so you can rub out any mistakes.

3 Why might a design engineer suggest using a ceramic material as a spacer to stop high voltage electric cables touching each other and their supports? (2 marks)

....

....

....

Hint

Look up the properties of ceramics and choose the best fit for your reason. Always write a sentence for your answer.

4 On the diagram clearly label the following:

a the anode

b the cathode

c the direction the electrons migrate in the electrolyte. (3 marks)

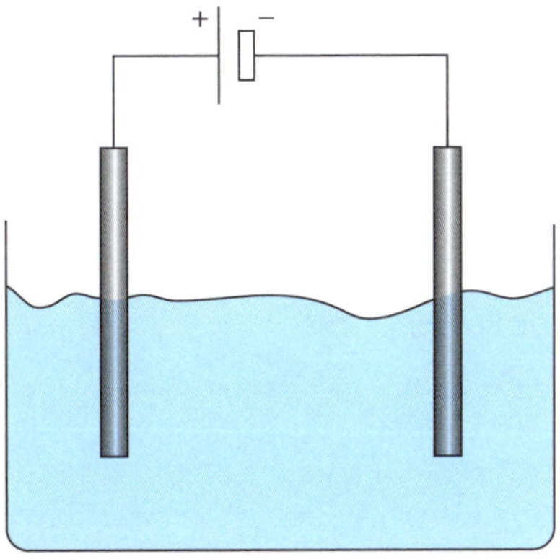

Hint

If you need help, there is a diagram in the textbook and also many similar ones online. Try to remember the answers before looking them up.

5 a What does GRP stand for?

A General rubber polymer ☐

B Gross rent potential ☐

C Glass reinforced polymer ☐

D Generation research program ☐

b Give one use for this material. (3 marks)

....

Hint

You need to know the answer to the first part of this question before you can answer the second part. Choose your use for the material carefully; don't try to be too clever. An obvious answer will give you just as many marks.

6 A sample of steel is found to have small particles of ceramic in its structure. What are these particles called in metallurgy? (1 mark)

I ....

Hint

Sometimes you might be asked for an answer that is hard to look up; at least here you have the first letter to help you research.

Long-answer exam-style practice questions

1 Rolling a metallic material can change the shape and properties of the material. The effects are different depending on whether the material was hot or cold when rolled. Explain the difference in the resultant structures and how this affects the properties of the material. (4 marks)

Plan your own answer

Your answer should cover these areas:

- structure
  - cold rolled – crystal structure compressed and bonds disrupted, larger crystals
  - hot rolled – crystal structure changed, smaller crystals
- properties
  - cold rolled – brittle and hard
  - hot rolled – ductile.

....

....

....

....

....

....

....

....

....

Sample answer

Comparing the differences between a sample of metal that has been hot rolled against one that has been cold rolled shows that the hot-rolled sample has smaller crystals and is less brittle, while the cold-rolled one has larger crystals that have been squashed and it is more brittle.

Comment

The candidate has understood the difference between the two processes but has not really explained why this affects the properties of the resultant materials. The 'squashed' crystals are broken and the bonds between the atoms are fractured, explaining why the sample is brittle. When discussing the hot-rolled sample, the candidate has used the phrase 'less brittle' but 'ductile' would show a better understanding of material properties. Saying 'compressed' instead of 'squashed' would also be more accurate.

Hint

Using the correct, technical terms shows a higher level of understanding. Now write your own answer to cover all the relevant points and using all the correct technical terms.

2 An architect wants to use timber for the beams and wall frames of a new house. As the structural engineer on the project, use your knowledge of the different properties of softwood, hardwood and engineered wood to suggest a suitable timber for the project. (6 marks)

Plan your own answer

Your answer should cover **one** of these areas:

- softwood, for example pine – cheap, easy to work, available, light, anisotropic
- hardwood, for example oak – strong, durable, in keeping with historic architecture, anisotropic
- engineered timber, for example Glulam – composite wood beam with excellent building properties, can be made in the correct shape for each application. Isotropic and strong in all directions.

....

....

....

....

....

....

....

....

....

....

....

Sample answer

The best solution to this building problem is a softwood such as pine. Pine is cheap, so this will keep costs down and it is easily available, so there won't be any supply problems. Softwoods are also easy to work, so joiners can work efficiently and light, so builders won't need any large lifting equipment. Softwood is better for the environment as softwood trees, such as pine, grow quickly and can be replaced easily.

Comment

This is a very good answer listing all the points that are relevant to the use of a softwood such as pine as a building material. A hardwood such as oak or an engineered timber such as Glulam could also be good answers, but with different advantages. This answer would be even better if the student used the technical term 'anisotropic' which applies to wood materials with a grain.

Hint

Timber is an engineering material but it is often overlooked. Engineers should understand the properties of timber as a material, as well as the more usual metals and polymers. Write your own answer for one of the other two options in the answer plan.

3 You are working as a marine engineer and you have been asked to design a new rudder blade for an ocean-going yacht.

You have the choice of steel or aluminium for the rudder-blade material. Using your knowledge of the corrosion process, describe the effect that seawater has on metallic components and choose the more suitable material. Explain your choice. Give two possible surface treatments that could reduce the problem and explain your reasoning. Suggest another material for the rudder blade that might be a better solution. (6 marks)

Plan your own answer

Your answer should cover these areas:

- aluminium – light, more resistant to corrosion
- steel – strong, corrodes easily
- aluminium is the better choice – strong enough but light, less likely to corrode
- surface treatment – paint or powder coating to protect the surface
- alternative material – composite materials, corrosion resistant, no surface treatment needed.

....

....

....

....

....

....

....

....

....

....

....

Sample answer

The best solution for a rudder-blade material for an ocean-going yacht is aluminium. Steel is stronger than aluminium, but aluminium is light and strong enough for the job. It is also less likely to corrode in seawater. Barnacles can also be a problem.

Aluminium can be painted or powder coated to protect it from the seawater and slow down the corrosion process.

Comment

This question would be easier to answer if the key aspects were highlighted. The candidate has answered *most* of the points, but they haven't demonstrated their 'knowledge of the corrosion process'. The examiner needs to see that the candidate understands the chemistry of the process and why aluminium is a better choice (an impervious layer of aluminium oxide on the surface resists corrosion).

The point about barnacles is interesting, but not relevant.

The candidate hasn't answered the last part at all. 'Suggest another material' is probably asking for a suggested composite material. This would only need cleaning and would not need painting, even in seawater.

4 Using two examples of your choosing, explain the difference in properties between crystalline and non-crystalline materials. (6 marks)

....

....

....

....

....

....

....

....

5 Describe the ageing processes that a polymer window frame might undergo when installed in the south-facing wall of a building. (6 marks)

....

....

....

....

....

....

....

....

6 Describe the three stages of creep in materials and how this might affect the material of an engine block. (6 marks)

....

....

....

....

....

....

....

....

....

7 Describe what happens to the structure of high-carbon steel when it is heated and then rapidly cooled. How does this affect the properties of the steel? Diagrams can be used to support your explanation. (6 marks)

....

....

....

....

....

....

....

....

8 Explain the reasons why the injection moulding process is used for many polymer components. (6 marks)

....

....

....

....

....

....

....

....

9 Discuss the process of welding and the effects it can have on the materials being welded. How can these effects be useful or detrimental? (6 marks)

....

....

....

....

....

....

....

....

Mechanical principles

Hint

The mechanical properties of a system describe the moving parts. How the parts move is governed by several different equations and principles, and affected by the materials used and the environment in which the system is found. This is a complicated topic but this section will help you practice the basics.

Recall activities

1 Match each of Newton's laws of motion to the correct description.

Law
First law
Second law
Third law

Description
The acceleration of an object depends on the mass of the object and the net force acting upon it
A body will remain at rest or will continue moving at a constant speed in a straight line, unless a force is acting on it
For every action or force, there is an equal and opposite reaction

2 Create a list of all nine sources of power stated in the textbook, and list at least one advantage and one disadvantage of each.

1 ....

2 ....

3 ....

4 ....

5 ....

6 ....

7 ....

8 ....

9 ....

3 Rub your hands together. Why do your palms get warm?

....

....

4 Explain what is meant by a torsion force.

....

....

5 Match each mechanical property to its description.

Mechanical property
Potential energy
Point load
Kinetic energy
Simply supported beam
Non-contact forces
Concurrent forces
Shear forces
Power

Description
Forces that act in a parallel direction to the surface of an object
Forces acting between objects that do not physically touch each other
Beam that rests on two supports and is free to move horizontally
Rate at which energy is transferred or converted
Forces where the lines of action all meet at the same point
Force applied at a single point along a beam
Energy that is stored by an object due to its position
Energy that an object possesses because of its motion

Short-answer exam-style practice questions

1 Which is the best description of the forces on an object moving at a constant speed? (1 mark)

A Unbalanced forces ☐

B Resultant forces ☐

C Balanced forces ☐

2 Which statement describes the relationship between acceleration and mass? (1 mark)

A Directly proportional ☐

B Inversely proportional ☐

C No relation ☐

3 A block of aluminium has a mass of 2.3 kg. What does it weigh on the Earth's surface? Assume the acceleration due to gravity is 9.81 m s^{-2} and give your answer to 2 decimal places. (2 marks)

....

4 A 20 kg rock and a 2 kg wooden log are dropped in a vacuum. The force of gravity on the rock is:

A 10 times as much as the force on the log ☐

B zero ☐

C the same as the force on the wooden log. (1 mark) ☐

5 A pallet of heavy machinery is sitting on a factory floor. It needs to be moved but there is no pallet truck or forklift available. Why would it be advantageous to put the pallet on wheels? (4 marks)

....

....

....

....

6 **a** An astronaut is testing their space suit at the aeronautical design centre. The suit has a mass of 60 kg and the astronaut has a mass of 80 kg. Assuming the acceleration due to gravity on Earth is $9.81\,m\,s^{-2}$, what does the astronaut weigh on Earth? (2 marks)

....

....

b The astronaut travels to Mars and puts on around 5 kg on the journey due to the high energy space rations they eat. Assuming the acceleration due to gravity on Mars is $3.71\,m\,s^{-2}$, what does the astronaut weigh on Mars? (2 marks)

....

....

Long-answer exam-style practice questions

1 An athlete is testing their strength by performing the push–pull sledge exercise. They push a weighted sledge along a high friction surface from one end of the gym to the other and then pull it all the way back to the beginning.

Describe the forces acting on the weighted sledge when travelling in both directions. (6 marks)

Plan your own answer

Your answer should cover these areas:

- weight
- gravity
- friction
- sliding
- mass
- push
- static

....

....

....

....

....

....

....

Sample answer

The athlete will brace themselves on the sledge to get enough grip to move forward. First, they will have to overcome the static friction of the sledge against the floor and then, once it is moving, the friction will decrease and the athlete will be able to get the sledge moving.

Comment

The 'plan' is just a list of key words, but the answer is well written and explains half the process well. However, the candidate hasn't discussed the return journey of pulling. Even though the same forces are involved in pulling, the question does ask for this and marks will be lost.

2 A compass usually has a two-ended pointer.

Explain why one end of the pointer always points north. What factors might stop the pointer moving freely? (6 marks)

Plan your own answer

Your answer should cover these areas:

- the pointer is magnetised
- attractive and repulsive magnetic forces cause the pointer to align with magnetic north
- a compass must be horizontal to reduce friction at the pivot.

....

....

....

....

....

....

Sample answer

The pointer on a magnetic compass is magnetised and one end will always point north and the other will point south 1. *If the compass is held at an angle, the pivot point will create friction and the pointer might not move very well* 2.

Comment

1 The first part of the answer doesn't mention the attractive and repulsive magnetic forces which, in the context of this exam, are very important. However, the second point **2** is relevant and will be worth marks.

3 Explain the difference between a renewable and a non-renewable energy source. Give two examples of each. (6 marks)

....

....

....

....

....

....

....

....

Hint

Think of the context of your exam when answering questions and ask yourself what knowledge the examiner is testing.

Hint

Look at the number of marks. There are probably two marks for the explanation and one mark for each of the four examples.

4 A bus has a mass of 3000 kg with just the driver on board. Calculate the acceleration of the bus when a force of 3000 N is applied to it.

The bus stops and takes on 20 passengers who weigh an average of 80 kg each. Calculate the acceleration of the bus with the same force applied.

Explain the difference. (6 marks)

....

....

....

....

....

....

....

5 Discuss the advantages of spreading a load uniformly across a beam when constructing a bridge. (6 marks)

> **Hint**
>
> The question mentions a bridge but all structures benefit from proper design to spread the load, so the context is not important.

....

....

....

....

....

....

6 Describe the process of annealing aluminium alloys. (6 marks)

....

....

....

....

....

7 Complete the table. One row has been done for you. (6 marks)

Material	Type	Example use
Stainless steel	Ferrous metal	Bolts and fixings
Carbon fibre		
Tungsten carbide		
Polythene		

Electrical and electronic principles

Recall activities

1 Label the three subatomic particles on the diagram of the atom.

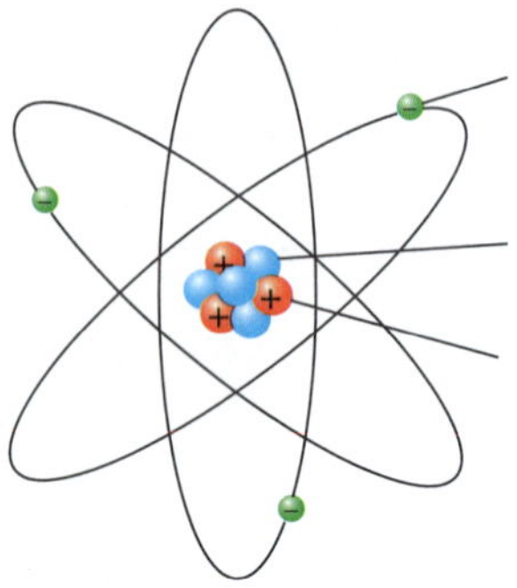

2 Give three examples of situations that use an electric motor.

1 ....

2 ....

3 ....

3 Match the parameter with the correct unit and then add the symbol for that unit.

Parameter
Capacitance
Inductance
Voltage
Magnetic flux
Power
Current
Resistance

Unit of measurement	Unit symbol
volt	
ampere	
ohm	
watt	
farad	
henry	
weber	

4 Fill in the triangles in the diagram with the correct symbols for Ohm's law and Watt's law.

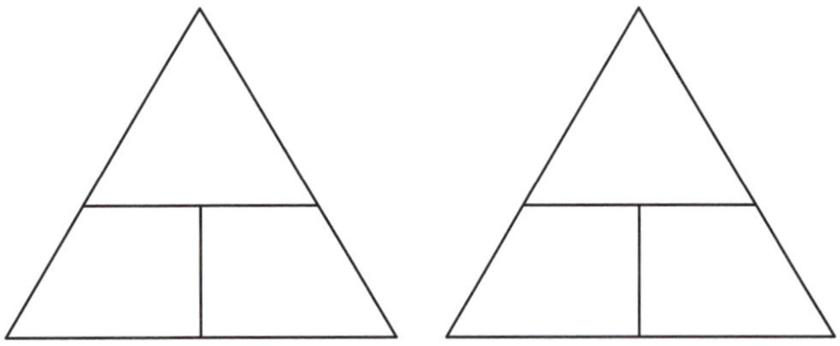

5 Consider the diagram. If R_1 is 100 Ω and R_2 is 200 Ω, what is the total resistance?

....

6 Which of these elements is a semiconductor?

Magnesium	**Carbon**	**Silicon**	**Iron**	**Silver**	**Hydrogen**

....

7 What sort of signal is shown in the diagram, analogue or digital?

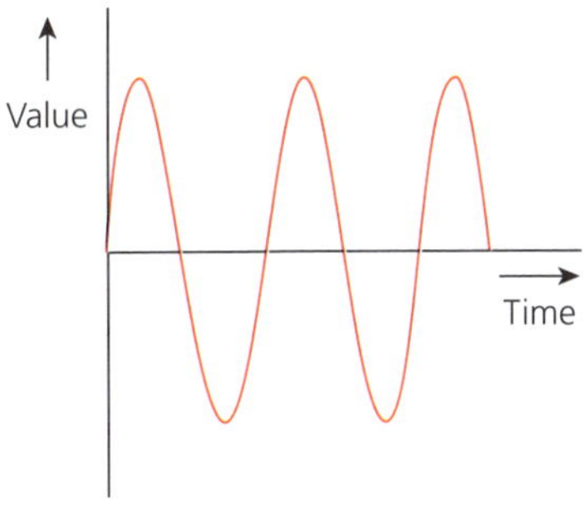

....

Short-answer exam-style practice questions

1 If R_1 is 300 Ω and R_2 is 200 Ω, what is the total resistance of the circuit in the diagram? (2 marks)

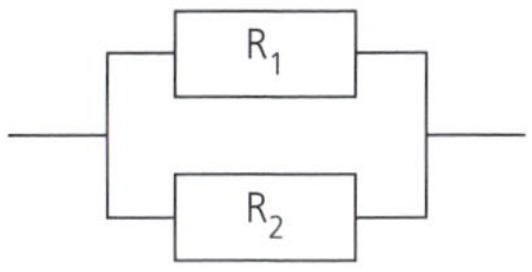

....

....

2 If C_1 is 10 µF and C_2 is 2.2 µF, what is the total capacitance of the circuit in the diagram? (2 marks)

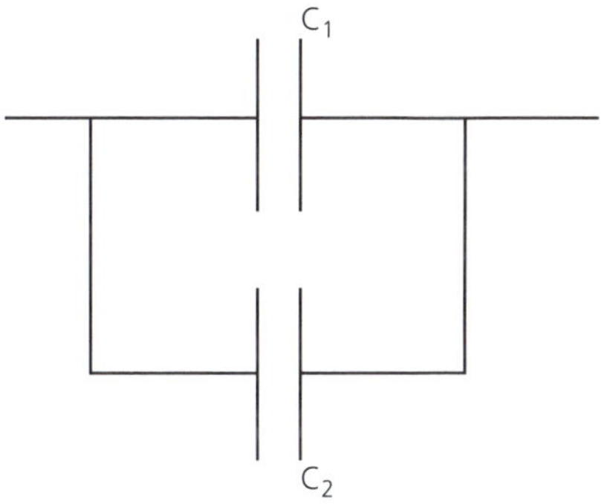

....

....

....

3 Explain why a system might be protected by a residual current device. (3 marks)

....

....

....

4 In this diagram of two analogue signals, what is the amount labelled $\theta°$ called? (2 marks)

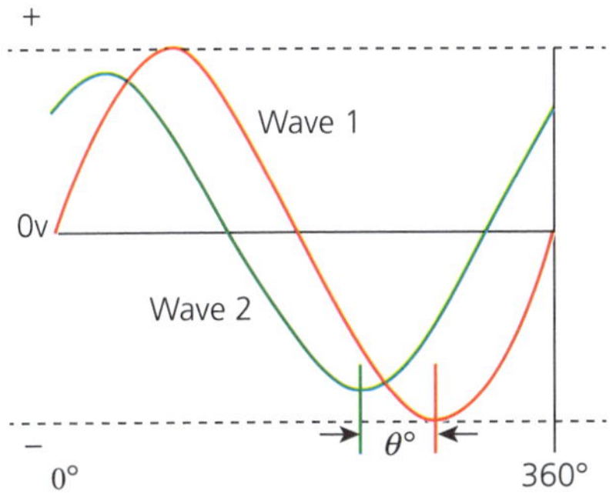

....

5 If V_s is 12 V and R_1 and R_2 are 2200 Ω, what is the output voltage at V_o? (2 marks)

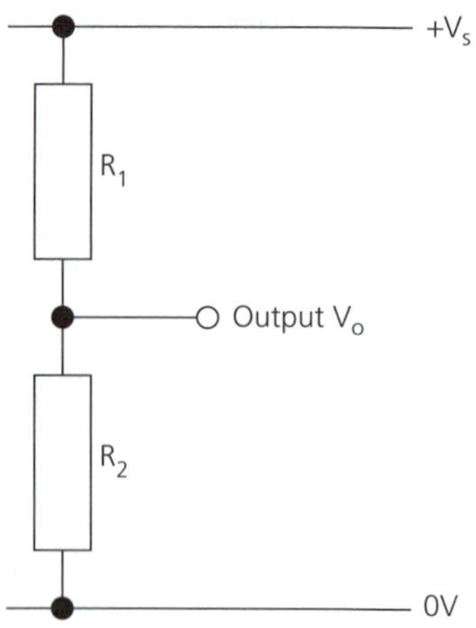

....

....

....

....

....

....

6 In the circuit in the diagram, how much power is being dissipated by R_2? (4 marks)

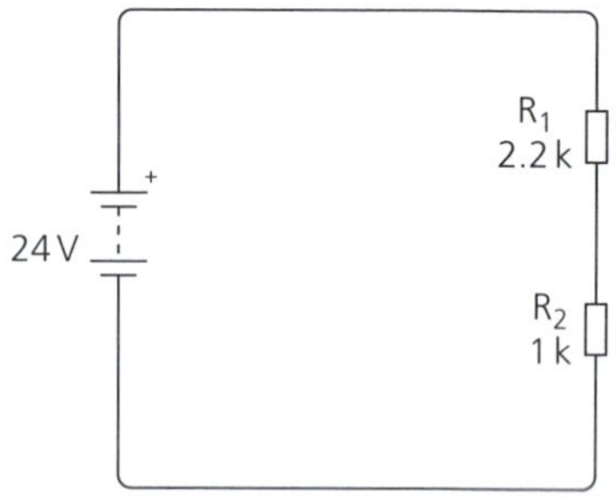

....

....

....

....

....

....

....

....

7 Calculate the total resistance of the resistor network between A and B shown in the diagram. (4 marks)

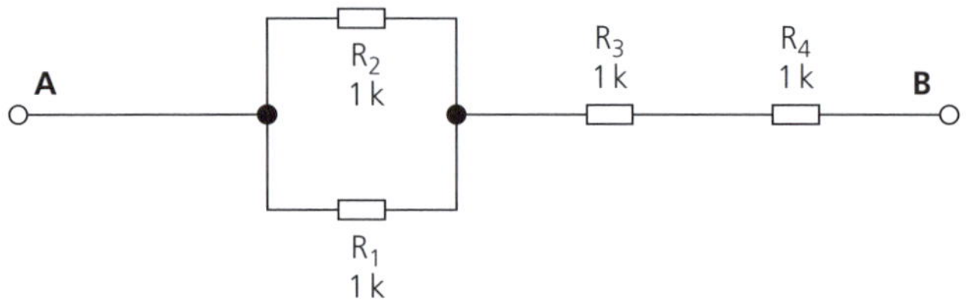

....

....

....

....

....

....

....

8 A solar panel is installed on a domestic garage.

a If the panel is producing 498 W at 14 V on a bright sunny day, what current is the solar panel producing? (2 marks)

....

....

b If the installer adds a further 11 panels to the same roof, how much power can the system generate on a similar day? (2 marks)

....

....

9 On the diagram of a magnet, identify the north and south poles and indicate the name of the lines. (4 marks)

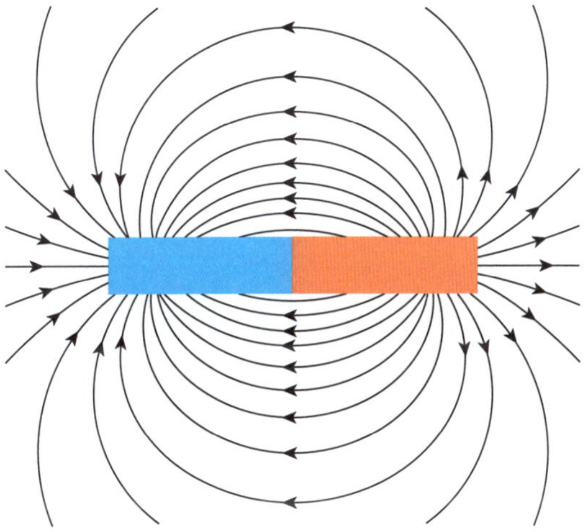

10 In the circuit diagram, the LED can be turned on by pressing the switch SW1.

a What voltage would be measured at Point A when the switch is pressed? (2 marks)

....

....

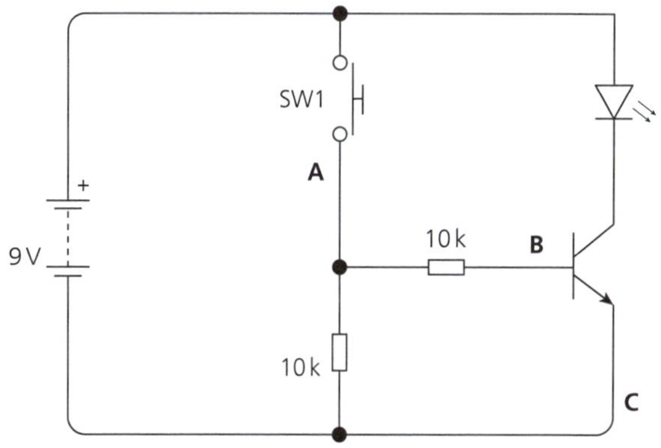

b What voltage would be measured between Points B and C when the switch is pressed? (2 marks)

....

....

Long-answer exam-style practice questions

1 Describe the difference between an analogue and a digital signal, and explain why a digital signal is less prone to interference than an analogue signal travelling the same distance at the same power. (6 marks)

Plan your own answer

Your answer should cover these areas:

- digital – on/off, square wave, filter interference
- analogue – varies from min to max, interference noticeable.

....

....

....

....

....

Sample answer

A digital signal can either be on or off and is usually shown as a square wave. This means that small changes to the signal caused by interference can be ignored or filtered out. With an analogue signal, the signal can vary anywhere between the maximum or minimum of the range of the signal and small alterations will be noticeable.

Comment

This is a good answer but it does not fully explain why digital signals are less prone to interference than analogue signals. Probably worth a maximum of 4 marks.

2 Explain one application of a diode in modern electrical/electronic engineering. (6 marks)

Plan your own answer

Your answer should cover these areas:

- **one** possible use from rectifier, voltage regulator, switch
- voltage regulator – stops voltages getting too high, protects circuitry.

....

....

....

....

....

....

Sample answer

A diode can be used as a circuit protection device because it limits the voltage in a circuit. It can also stop electricity flowing backwards in the circuit and harming circuit components because a diode only lets current flow one way.

Comment

This is a good answer. To practice, choose another use for diodes, such as in rectifiers, as voltage regulators or as switches, or one of your own choosing and describe how it works.

3 What are the advantages of hierarchical design in electrical engineering? (6 marks)

....

....

....

....

....

4 Describe why a diode junction might break down. (6 marks)

....

....

....

....

....

5 Describe three ways a technician might use a multimeter. (6 marks)

....

....

....

....

....

6 Describe some of the challenges facing the designer of a new electricity pylon. (6 marks)

....

....

....

....

....

Mechatronics

Recall activities

1 What does PLC stand for?

....

2 Complete the table with examples of these lever types.

Order	Example
First order	
Second order	
Third order	

3 List as many advantages of using robots in manufacturing as you can think of. Now list as many disadvantages as you can.

....

....

....

....

4 Match each sensor name with the correct sensed parameter.

Sensor name
Thermistor
LDR
Potentiometer
Reed switch

Parameter
Rotational position
Temperature
Magnetism
Light level

5 Find a pedal bicycle and challenge yourself to identify as many mechanisms as you can.

....

....

....

....

6 Find a real-life example of signal conditioning and write an explanation.

....

....

....

7 In as few words as possible but keeping the meaning clear, explain the difference between pneumatic and hydraulic systems.

....

....

8 On a separate piece of paper, create a safety poster that explains the hazards of using compressed air in the college workshop.

9 What is the name of the linkage type shown in the diagram on the right?

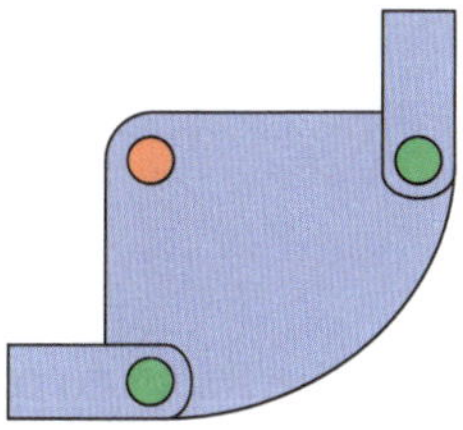

....

10 Investigate the meaning of the word 'cobot' and think how it could apply to an industry you know about.

....

....

Short-answer exam-style practice questions

1 The diagram on the right shows part of the pinout of a microcontroller. State what each of the pin labels might mean. (3 marks)

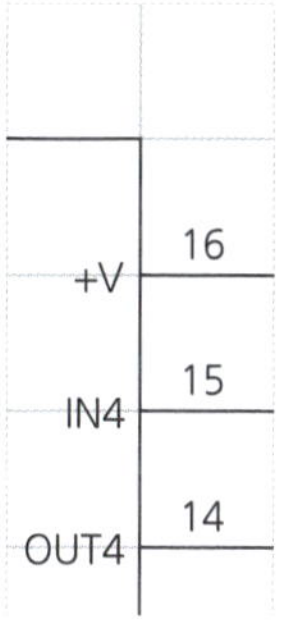

a +V ....

b IN4 ....

c OUT4 ....

2 A gear train has a driver gear with 25 teeth and a driven gear of five teeth. What is the gear ratio? (2 marks)

....

....

3 The diagram shows a cam system. Label the four parts using words from the list. (4 marks)

Camshaft	**Cam**	**Crank**	**Follower**

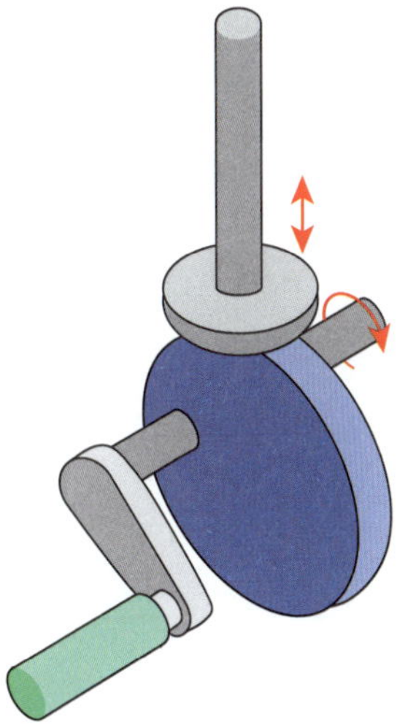

4 Give an example of a second order lever. (2 marks)

....

5 Automated production installations increasingly use artificial intelligence (AI). Explain how AI will change the use of this type of equipment in industry. (4 marks)

....

....

....

....

6 The input to a two-gear gearbox is 400 revolutions per minute (rpm) and the output is 1200 rpm. If the input gear has 90 teeth, how many teeth does the output gear have? (2 marks)

....

....

Long-answer exam-style practice questions

1 The diagram shows how a PLC might interact with input and output devices in a control system.

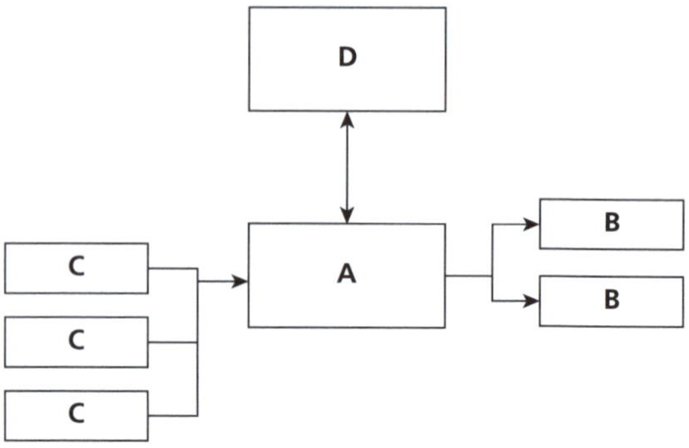

a State what each block might represent and give examples. (3 marks)

A ....

....

B ....

....

C ....

....

b If D is the user interface console, what functions might be performed at this station? (3 marks)

> **Plan your own answer**
>
> Your answer should cover these areas:
>
> - A – PLC microcontroller; B – outputs; C – inputs
> - D user interface – programming and monitoring.

....

....

....

Sample answer

a A Microcontroller, PLC

B Outputs, servos etc

C Inputs, sensors etc

b The user interface could be used for programming the PLC and monitoring the way it is working.

Comment

The candidate has answered the question using the planning notes. However, the use of 'etc' is not recommended in an exam – an examiner will give no marks for information that they have to think of themselves! More examples of input and output devices will convince the examiner that you know what you are talking about. Including some information about what action an operator might take in response to the monitoring information would improve the answer. If the equipment needs maintenance, what would they do? What if the equipment is faulty and needs an emergency stop?

Write your own complete answer that includes all of this extra information.

2 You are a production design engineer in a factory with many different production lines for both mass-produced items and one-off bespoke products. Your manager asks you to investigate whether installing PLC equipment would help production efficiency. Discuss the advantages and disadvantages of using robots on large and small production lines. (6 marks)

Hint

If you have completed Question 3 in the Recall activities section, you should already have the basis of good plan for this answer. The emphasis in this question is on applying the advantages and disadvantages to different *scales* of production.

Plan your own answer

Your answer should cover these areas:

- large scale
  - advantages – fast, reliable, accurate, repetitive tasks, 24/7
  - disadvantages – expensive to install, highly trained technicians and engineers needed for installation and maintenance
- small scale
  - disadvantages – set up takes time so not efficient for small scale, inflexible.

....

....

....

....

....

....

....

....

Sample answer

The main advantages of investing in robotic or PLC equipment are the speed and accuracy of robotic equipment. This would be of great use on a mass production line as robots are good for repetitive tasks and do not lose focus or accuracy. They can run 24/7, so the production line can run all day and night. In small scale production where the product might be one-off or 'bespoke', humans could be better for the work. They can be flexible and use their brains to make decisions about how to work.

Comment

The first part of this answer is good, clear and correct. There could have been more discussion of the advantages of robotic control, but the answer would be worth three marks.

However, the candidate has misunderstood the second part of the question. The question asks for the *disadvantages* of robots in a small scale production, not the advantages of using human operatives.

Write your own answer and change the emphasis of the second part.

3 In the pulley system in the diagram, pulley 1 is the same size as pulley 2.

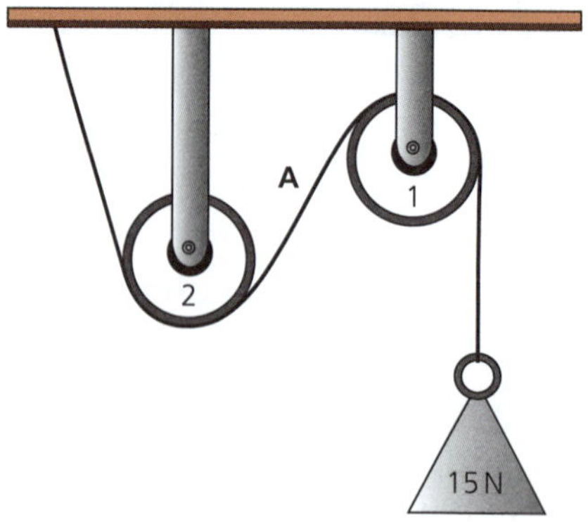

a What force is required in the rope labelled A to raise the load off the ground? Give your answer in newtons (N). (2 marks)

....

b Give an example of a pulley system in industry or construction and explain how it works to the advantage of the user. (4 marks)

....

....

....

....

4 Explain why a manufacturer of breakfast pastries would use pneumatic actuators in the production control systems, instead of electric or hydraulic actuators. (6 marks)

....

....

....

....

....

....

Engineering and manufacturing control systems

Recall activities

1 Fill in the truth tables.

NOT	I/P	O/P
	0	
	1	

AND	I/P1	I/P2	O/P
	0	0	
	1	0	
	1	0	
	1	1	

OR	I/P1	I/P2	O/P
	0	0	
	1	0	
	1	0	
	1	1	

2 Find the key terms in this wordsearch using these clues.

A	R	O	T	A	R	A	P	M	O	C	J	N	Q	O
Q	X	L	W	Y	Y	Y	S	E	G	F	F	R	U	M
A	W	Q	Q	F	K	W	Z	U	H	V	W	F	U	O
F	J	Z	K	E	J	I	Q	A	D	K	Q	E	D	K
N	K	P	S	E	N	S	O	R	G	C	E	O	D	T
P	C	A	Q	D	I	A	O	D	Z	O	E	K	V	T
B	P	L	N	B	W	O	P	T	U	L	I	Q	F	A
Q	P	L	O	A	I	X	S	G	M	B	N	I	C	Y
W	T	W	C	C	L	L	E	J	T	M	P	T	Z	T
M	Z	W	H	K	V	O	J	N	M	K	U	N	Q	M
R	D	Q	M	T	S	A	G	D	E	A	T	G	M	I
A	G	N	Y	I	M	D	I	U	T	M	L	J	T	O
V	Z	K	I	M	U	F	R	O	E	A	H	Q	G	Q
U	U	U	Z	E	S	S	R	Z	H	J	H	K	L	X
C	A	O	O	R	S	Y	E	S	A	X	B	R	Z	L

a Signal received from the environment into a system

b Device for changing the state of a signal for a set period

c Device for comparing two signals

d Device that turns a signal into movement

e Temperature is an .... signal

f Part of a signal that returns to affect the input

g A system diagram that uses rectangular shapes is a .... diagram

h A device that measures an input signal

3 Match each medium used for control to the name of the application.

Medium
Electronic
Pneumatic
Hydraulic

Application
Compressed air or vacuum pressure
Fluid movement
Electricity

Short-answer exam-style practice questions

1 Explain why a control system might use analogue sensors to measure some parameters and digital sensors for others. Give one example of each type of sensor. (4 marks)

....

....

....

....

Sample answer

A control system might use various types of sensor to detect parameters in a system. Analogue sensors measure a range of values from a minimum to a maximum, and anything in between. An example of this would be temperature, where the value could be in the range 30°C to 60°C, but also anything in between. Another example is light level which can be any value.

A digital signal is either on or off, so this could be whether a product is in the correct position or not.

Comment

This example answer is good enough for full marks. However, the candidate has given *two* examples of analogue sensors (temperature and light) when only one was required. If one example had been wrong, the examiner might assume the candidate didn't know what an analogue sensor is. Do not 'over answer' a question unless you are very confident.

Long-answer exam-style practice questions

1 You are designing a production plant for the manufacture of oat cookies. During the manufacturing process substandard cookies are removed from the conveyor by a quality control robot arm. Explain why you might use pneumatic actuators for this arm. Explore any alternatives and explain why they might be less suitable. (6 marks)

Plan your own answer

Your answer should cover these areas:

- pneumatic actuators – food stuffs require minimum contamination, use compressed air, fast moving, accurate, low power usage, reliable
- hydraulic actuators – not heavy work, so hydraulics are not required, could leak fluid
- electronic solenoids

Hint

Asking questions when you are out in industry is useful to broaden your knowledge. Never be afraid to ask questions. Now write your own answer addressing the option of electronic actuators more clearly.

Sample answer

In the manufacture of food stuffs everything must kept clean and free of contamination. Pneumatic actuators will not drop oil or give off vapours, apart from compressed air. They are also fast moving, accurate, have a low power usage and are quite reliable so they would be good for 24hr working. The alternative would be electronic solenoids or hydraulic actuators which would be not necessary as the loads would be small.

Comment

This is quite a full answer, apart from the fact that, in the last sentence, electronic actuators are mentioned but the reasons for using them or not are not covered. Their advantages and disadvantages for this type of use in the food industry should be made clear.

2 Give an example of a closed loop system, draw the block diagram for the system and explain how each part of the system works. Include suggested sensors and output components that might be used in the system. (6 marks)

Quality management

Recall activities

1 Match the key terms to the definitions.

Term
Technical specification
Compliance
Tolerance
Competence
Legislation
Testing
SOPs
Assurance
Conformity
Methodology
Sampling

Description
A law or set of laws passed by Parliament
Certainty about something
Compliance with standards, rules or laws
The process of selecting batches of products for testing
Being in accordance with commands, rules or requests
A system of methods or rules used for a particular activity
The ability to do something effectively, with sufficient knowledge, judgement and skill
Documents that define the requirements for projects, products or systems
Allowable amounts of variation for a specified quantity
The process of assessing product sample batches for conformity
Step-by-step instructions for routine operations

2 DMAIC is a well-known acronym for define, measure, analyse, improve and control. Make up a mnemonic to help you remember this and complete the table.

Letter	Memorable word
D	
M	
A	
I	
C	

Short-answer exam-style practice questions

1 A company samples its product at the end of the production line and rejects any imperfect items. It does not check its manufacturing process to stop this happening. Is this quality control or quality assurance? Explain why you have given that answer. (3 marks)

....

....

....

Sample answer

Just throwing away products that are no good will not help the company solve its problems. It will need to use proper quality control and quality assurance techniques to stop this happening.

Comment

The candidate hasn't really answered the question. First, they should identify whether the example given is quality control or quality assurance – it is quality control. Customers will not receive imperfect items, but the company will still have them rolling off the production line, wasting money and time. Write your own answer.

2 Explain why a company might try to achieve ISO9001 accreditation. (3 marks)

....

....

....

....

....

Hint

Having ISO9001 accreditation means the company can market itself as a business that meets particular quality criteria, so customers can be sure the company is properly run and will be a reliable supplier.

Long-answer exam-style practice questions

1 Explain how total quality management (TQM) can be applied in a company and how it will help the company work towards more efficient quality management. (6 marks)

Plan your own answer

Your answer should cover these areas:

- definition of TQM
- aims of TQM – detect and reduce errors in manufacturing, improve supply-chain management, enhance customer experience, ensure employees are adequately trained
- state that each aim is checked at the four stages of TQM – plan, do, check, act.

Hint

You might not know what TQM is, but you should be able to put together a reasonable discussion about quality control and assurance, which would gain some marks. No answer always gives zero marks.

Now write your own answer without using bullet points.

....

....

....

....

....

....

....

Sample answer

Total quality management (TQM) is a method of looking at the way a company works and how it can improve the way it works to be more efficient and reduce defects in manufacturing. The main aims are to:

- *detect errors in manufacturing on the production line*
- *make sure the materials used in production are correct for the purpose*
- *make sure all machinery is used correctly*
- *make sure all staff are trained in what they need to know and do*
- *check all resources coming into the factory are to the right specification*
- *ensure that the customers are happy with the product.*

To do this, the company will look at each of these points and use the 'plan, do, check, act' process to produce improvements.

Comment

The candidate has answered all the points that are needed for a good mark. One thing to note, however, is that if this were a quality of written communication (QWC) question, the use of bullet points might count against the candidate.

2 A company makes copper tubing. Discuss three points in the production cycle at which the company might use sampling for quality control. For each point describe what methods, tools and techniques could be used. (6 marks)

....

....

....

....

....

....

3 A company making enclosures for underwater pumps has a new client. This client would like the enclosures to be a different size to the ones the company usually makes, so they can fit into a rescue vehicle's equipment bay.

Explain what steps the company might take to ensure that all documentation is controlled, correct and future proof. (6 marks)

....

....

....

....

....

....

Hint

It doesn't really matter what the product is, any company that designs and makes a product should be keeping all documents up to date and secure. Consider a variety of different types of business documents, for example technical drawings, contracts, orders, emails and financial statements.

Health and safety principles and coverage

Hint

Some people think health and safety stops them 'getting the job done', but this dangerous mindset causes accidents and even deaths. UK governments have passed several Acts of Parliament to keep everyone safe and to protect our health.

The best way to revise the topic is to read the regulations and think about how they apply in real life situations. Look for health and safety rules when you are on work placement.

Recall activities

1 Follow the link www.hse.gov.uk/simple-health-safety/risk. Now review your knowledge by writing a short note to your manager (tutor) about the current state of health and safety in the workshop.

....

....

....

....

....

....

2 Are the following statements true or false?

- The employee is responsible for health and safety in the workplace. ....
- It is optional for employees to wear PPE. ....
- Good 'housekeeping' is important to prevent accidents. ....
- Employees should report health and safety risks they see in the workplace. ....
- The employer is responsible for health and safety in the workplace. ....
- A hazard is defined as anything with the potential to cause harm. ....
- The risk is the likelihood of something happening. ....
- Wearing the correct PPE is more important than eliminating the risks. ....
- When a proposed activity is identified as 'high risk', a 'permit to work' should be produced. ....
- LOTO stands for 'lock out, tag over'. ....
- Health and safety at work extends to consideration of the environment. ....

3 The acronym COSHH stands for 'control of substances hazardous to health'.

Make up a mnemonic to help you remember this acronym and complete the table.

Letter	Memorable word
C	
O	
S	
H	
H	

4 Find the posters at college that display the Health and Safety at Work Act. Where are they?

....

....

....

....

5 What does RIDDOR stand for?

....

6 It is the responsibility of an employer to provide health and safety training to their employees. When should employees be given training?

A When they are given new responsibilities at work ☐

B When they first join the business ☐

C At regular intervals during their employment ☐

D All of the above ☐

7 Arrange these steps in a risk assessment. Write the numbers 1–4 in the boxes to indicate the correct order.

a Identify who might be harmed ☐

b Decide how the level of risk may be controlled ☐

c Identify any hazards ☐

d Evaluate the risks ☐

8 How many employees does a company have to have before the Management of Health and Safety at Work Regulations (MHSWR) 1999 come into force?

....

9 Find out what these abbreviations stand for and write them in the table.

Abbreviation	Full form
HASAWA	
MHOR	
RIDDOR	
DSEAR	
COSHH	
PPE	
LOLER	
CEMFAW	
SSoW	
HSE	
HAZOP	
LOTO	
MHSWR	
HAZID	
MSDS	
PUWER	
WHAR	

10 Match the key terms to the definitions.

Term
Asphyxiation
Duty holders
Accident
Legislation
Regulations

Definition
A set of rules to be followed
An unplanned and unexpected occurrence
A set of rules and regulations passed in Law
Not having enough oxygen to breath
Someone with the responsibility for health and safety

Short-answer exam-style practice questions

1 Why would a company provide PPE for its employees instead of letting them buy their own? (4 marks)

....

....

....

....

2 When would the use of a mechanical lifting aid be beneficial? (2 marks)

....

....

....

Sample answer

When the operator cannot lift the heavy load without strain or possible injury.

Comment

The candidate has mentioned two good reasons so full marks are awarded.

3 This is part of a risk assessment for soldering in a college setting.

Hazard	Who/what might be harmed and how	Existing control measures	Risk rating		
			Likelihood/5	**Consequence/5**	**Rating/25**
Solder spitting in eye	Operator hurt/ blinded	Operator supplied with goggles	3	4	
Burning hand on soldering iron	Operator hurt/burnt	Operator given adequate training	2	3	
Heat melts sleeving on component	Component damaged or broken	Operator given adequate training	2	1	

a Complete the rating column of the table.

Hint

If you don't already know, look up how to calculate the rating column. Spotting the most urgent risk is straightforward once you have completed the calculations. Risk assessments should always prioritise the possible harm to humans, so remember that when answering part c.

b Which is the most urgent risk that has been identified?

....

c Why is the consequence of melting the sleeving rated so low when it might break the product? (4 marks)

....

....

....

....

4 Why is stress a factor in health and safety? (3 marks)

....

....

....

....

....

Hint

Think of how employee stress has an impact on the company, as well as on the employees themselves. How would an operator under stress work, compared to one that was relaxed and confident?

5 When might an inspector from the HSE (Health and Safety Executive) ask to look round a company's premises? (4 marks)

....

....

....

....

6 Give two examples of regulations that control health and safety in engineering. (2 marks)

....

....

....

Hint

Even though you won't have your textbook in the exam with you, looking these facts up now will help you remember them later.

Long-answer exam-style practice questions

1 A company is sending all its waste to landfill. This includes mixed packaging waste, rejected products and broken electrical parts. What could they do to improve their practices with respect to the environment? (6 marks)

Plan your own answer

Your answer should cover these areas:

- reducing waste in the first place
- sorting waste
- recycling waste correctly
- reusing waste where possible
- sending electrical waste to a waste electrical and electronic equipment (WEEE) recycler.

....

....

....

....

....

....

2 Explain the work of the Health and Safety Executive in the UK. (6 marks)

Plan your own answer

Your answer should cover these areas:

- HSE – keeps workers safe all over the UK, offers guidance on regulations, checks compliance with legislation, prosecutes when needed.

....

....

....

....

....

....

....

....

....

....

Hint

Although this seems like a simple question, there are six marks to gain, so a simple answer won't do. Answer it yourself and *then* mark it to see if you have six clear and relevant points to gain marks.

Sample answer

The Health and Safety Executive (HSE) is the organisation that looks after health and safety in the workplace. It applies everywhere in the country and it was designed to keep workers safe in work. The HSE has inspectors that can come and inspect any company or workplace whenever they think it should be done. They will check that all the health and safety rules are being followed by the company and prosecute them when they haven't followed the rules.

Comment

This is quite a good answer but it doesn't cover the way the HSE supports companies by giving clear guidance about up-to-date requirements. In your answer make sure you address these points.

3 The use of personal protective equipment (PPE) is often mandatory in an engineering environment. Discuss what this means and why a company might have rules about using PPE. Give three examples of the types of PPE that might be covered by a company's rules. (8 marks)

Plan your own answer

Your answer should cover these areas:

- PPE – protects workers from harm
- mandatory – clear enforceable rules, legal responsibilities
- examples – eye protection, ear protection, overalls, goggles.

....

....

....

....

....

....

....

....

....

....

Sample answer

Using PPE can help someone keep safe when they are working. Using ear defenders can stop them being deafened, eye protection can stop them being blinded, and barriers between them and dangerous machinery can stop them getting hurt.

Comment

Unfortunately, the candidate has missed the point of the question. It is always worth using a highlighter for high-mark questions. Here is the question again but with highlighter marking to show the important parts that should be covered in the answer.

> The use of personal protective equipment (PPE) is often mandatory in an engineering environment. **Discuss** what this means and why a company might have **rules** about using PPE. Give **three examples** of the types of PPE that might be covered by a company's rules.

So, has the candidate answered the question asking *why* the company has rules about wearing PPE? If a company makes wearing PPE mandatory (compulsory), they can ensure everybody follows the rules and take action if anyone refuses. This means the company is keeping employees safe and complying with health and safety laws.

Only two of the three examples given are PPE (ear defenders and eye protection). While using barriers as a safety precaution is a good idea, it's not personal protective equipment.

Write your own answer.

4 A company buys a new piece of machinery that, when installed and working, produces a loud and high-pitched squealing noise. The noise is normal for this type of machinery. Outline what health and safety measures the company should put in place to protect the operators of this machinery and anyone who may work in the environment. (6 marks)

Hint

Use a highlighter on the following questions before you try to answer them.

....

....

....

....

....

....

....

....

Hint

PPE, signage and training are the three most important things here. What do you think the company should do if their control measures don't work? Will the operators need any support in their work such as rest breaks or regular hearing checks?

5 A small start-up company is going to make specialist exhausts for automobiles. They will be working with pipe cutters and bending machines, and will be welding parts together.

a What PPE would you suggest the company invest in for their employees?

b If the employees are going to be working on vehicles raised on a four-post lift, what extra precautions should the employers take? (6 marks)

Hint

Think of what precautions you use for these tasks in the college workshop and use that as a guide. Never forget that training is the biggest preventer of accidents.

....

....

....

....

....

....

....

....

6 Explain the steps a company should follow to assess and reduce risk in the workplace. (8 marks)

Hint

The textbook has many examples and descriptions of steps that a company should take. Read page 193 onwards, then write your answer. Keep looking back to the question to check you are really answering what it asks.

7 Why would a company make rest breaks and a factory shutdown part of their working practices when these reduce production time? (8 marks*)

Hint

Think about stress – do you work best without rests or holidays? What are the advantages to a company in having contented employees?

Business, commercial and financial awareness

Recall activities

1 Choose whether these business costs are direct or indirect costs. Tick the appropriate column and then check your answers.

Cost to the company	Direct	Indirect
Equipment or tools needed to make a product		
Raw materials used to make a product		
Rent for a factory making more than one product		
Cost of selling a product		
Maintenance		
Wages of the catering and cleaning staff		
Insurance		

2 Match the key terms to the definitions.

Term
Procurement
Turnover
Stock forms
Shares
Clauses
Accountant
Patent
Depreciate
Contractor
Bonuses
Income

Definition
Clearly defined sections of a contract or other legal document
Reduce in value over time
Total amount of money received by a company from the sale of goods or services over a certain period of time
A legal document that grants an owner exclusive rights over an invention
The process of identifying and buying materials, tools and equipment for use by a company
Units of ownership of a company that can be bought and sold
Money that comes into a company
Someone who is self-employed but agrees to complete work for a company for a set fee
A qualified person who prepares and audits a company's financial records
Standard shapes and sizes in which materials are available
Extra money over and above the salary or wages of an employee

3 Find out the meaning of the acronym 'TIMWOODS' and, on a separate piece of paper, create a graphic to help remember it.

Short-answer exam-style practice questions

1 A sales department decides to create an online survey which is sent to all contacts. What is this survey an example of? (1 mark)

- A User questioning ☐
- B Market research ☐
- C Canvassing ☐
- D Scatter enquiry ☐

2 How might a bicycle manufacturer be said to be a value added company? (3 marks)

....

....

....

Long-answer exam-style practice questions

1 You are the manager of a newly established company that uses injection moulding to make high quality polymer (plastic) parts. Another company on the same industrial estate has a new project in development that needs a supply of small plastic parts. What commercial points would you need to consider when drawing up a tender for the work? (5 marks)

Plan your own answer

Your answer could cover these areas:

- customer is close by so reduced shipping costs
- local company means good communication
- quantity that can be produced
- quality criteria
- timescale for supply
- potential to produce greater quantities
- potential for negotiation on costs.

Hint

The key terms in this question have been **highlighted like this**. You may spot other words you'd like to highlight. It is always a good idea to use a highlighter in exams to help you focus on the questions.

....

....

....

....

....

....

....

....

Sample answer

The manager of a newly established company would need to secure new business wherever it can be found. A local customer would have the advantage of shipping costs being considerably reduced. There would be no need for large trucks or container shipments. A small van might be sufficient for delivery, even if it makes several trips. This will reduce the price that can be offered to the customer and increase the chance of winning the contract.

The product is high quality and this could be used as a good marketing strategy. Quality could be checked all the way through production and before the product is sold. Checking the quality of materials, the tools and machinery and the final product all help achieve good quality.

If the company gets the contract, it could lead to bigger things as sales increase.

Comment

The candidate mentioned delivery costs but not cost of production and did not touch on timescales. There was quite a bit about quality, but this is mostly irrelevant to this question about tendering. Write your own answer that covers five points from the plan.

2 Discuss the work of three departments that are often found in engineering companies: sales and marketing, finance, and research and development. Explain how these three departments interact with each other. (6 marks)

Hint

This question is asking you to define the functions of each department and to say how they interact. There could be one mark for each department definition, and one mark for each of three interactions, making six marks all together.

....

....

....

....

....

....

....

....

....

....

....

....

Professional responsibilities, attitudes and behaviours

Recall activities

1 The table shows attitudes and actions in the workplace. Tick the columns to show which you think apply to the employee, the employer or both.

Attitudes and actions in the workplace	Employee	Employer	Both
Being punctual			
Paying wages/salary on time			
Setting tasks that are suitable for a person's capabilities			
Acting professionally at all times			
Providing training where needed			
Completing work as requested			
Providing rest rooms and breaks for meals			
Treating fellow workers with respect			

2 What does the acronym CPD stand for?

A Collaborating personnel department ☐

B Continuous professional development ☐

C Cataloguing production designs ☐

D Cross-production directions ☐

Short-answer exam-style practice questions

1 Give two reasons why a member of staff might make 'human errors'. Describe how the company can help avoid the errors you have described. (4 marks)

....

....

....

....

....

....

Hint

Check the textbook for two causes of human errors and explain how they affect staff. Make sure you revise all the reasons for workplace stress.

Long-answer exam-style practice questions

1 Why might an engineer apply for chartered status and why would their employer support them? (6 marks)

Plan your own answer

Your answer should cover these areas:

- for the engineer – professional recognition, access to CPD, networking opportunities, promotion
- for the employer – staff motivation, company reputation, effective CPD.

Hint

As you can see here the planning is almost the same length as the final answer but the answer is written in a more coherent style.

Sample answer

The advantages of **professional recognition** *and chartered status to the engineer would be* **professional recognition** *and access to good quality CPD sessions, a chance to network with other engineers and to improve their chances of promotion or career progression. The advantages for the company of supporting the engineer in this would be a more motivated staff member and a better reputation when recruiting. The professional bodies might also provide good value CPD opportunities.*

Comment

The only mistake here is the repetition of professional recognition, as highlighted, but this is easily corrected reading through the answer. Always read through your answers.

2 What are the main responsibilities of a company when recruiting new staff? (6 marks*)

Stock and asset management

Recall activities

1 Review the different types of stock in section 15.1 of your textbook. Match the 'stock type' to the 'items held' for a company that makes vacuum cleaners.

Stock type
Raw materials
Subassemblies
Components
Consumables
Work-in-progress
Fasteners and fixings

Items held in stock
Switches
Epoxy-resin adhesive
Self-tapping screws
Motor units
Plastic granules
Half-built vacuum cleaners

2 Fill in the blanks in the paragraph below using terms from the box.

Warehouse **Stock** **Obsolete** **Depreciation** **Stocktake**

A company manufacturing telescopes has a lot of out-of-date .... stock in its .... It decides to complete a .... to see if any products can be sold or recycled. Any items that are kept will reduce in value because of ....

Short-answer exam-style practice questions

1 Why do companies use barcodes to label their stock? (2 marks)

....

....

2 A stock item depreciates at 20% of its original value per year. How long will it take to 'write off' the item in stock? (2 marks)

....

....

3 Which of these pieces of information might be contained in a catalogue of parts? (1 mark)

A Name of part ☐

B Date purchased ☐

C Supplier ☐

D All of the above ☐

Long-answer exam-style practice questions

1 Discuss how obsolete stock can cost a company money. Give examples where appropriate. (6 marks)

Hint

Giving examples is very useful to show you understand the topic. Use examples from your own experience where you can.

Sample answer

Stock has to be stored somewhere and using a building or room could mean paying rent, heating and lighting. Some stock can be become worthless if kept too long. Stock needs to be counted in a stocktake and entered into the company's database or catalogue. This means staff will spend time counting and doing admin for stock that has no value to the company.

Comment

This is pretty good answer but will it get full marks? Probably not, as the questions asks for examples. The candidate should give at least two examples of stock that has cost a company money. Write your own answer that includes examples.

2 Why is it more expensive to store frozen food than screws and bolts? (6 marks)

Hint

You may not have worked in or studied a food manufacturing environment, but this question can be answered using general knowledge. Where is frozen food stored? Is this storage free? Always tailor your answer for the work environment. So, say 'storing frozen food is more expensive because …' instead of 'I keep my food in a freezer so it is more expensive because …'.

3 When and why might a company make products to be kept in stock? (6 marks)

....

....

....

....

....

....

4 Discuss one example of the use of robotics in stock control or movement, and explain why a manufacturer might find this useful. (6 marks)

> **Hint**
>
> If you get an open question that asks you to give an example, use one that you really understand. Do not use an exam as a chance to show off.

....

....

....

....

....

....

....

5 Explain how and why a company might use pick-to-light in their assembly department. (8 marks*)

> **Hint**
>
> More marks mean more information is required in your answer. This question is also marked with an asterisk, meaning it is a QWC question, so your answer should be planned carefully, then written out in full in good English.

....

....

....

....

....

....

....

....

....

....

....

Continuous improvement

Recall activity

1 Match the key terms to the definitions.

Term
Lean manufacturing
Kanban
Total cycle time
Value stream mapping (VSM)
6S
Throughput
Kaizen
Key performance indicators (KPIs)

Definition
Continuous improvement through small, incremental positive changes
Quantifiable measures used to evaluate the success of an activity or organisation
Analysing how materials and information flow through a company
Number of products manufactured per unit of time
Manufacturing with the minimum waste
Managing the flow of work in a manufacturing company by using instruction cards or boards
Time to manufacture a customer order from start to finish
One approach that is used to help create an organised, efficient and safe workplace

Short-answer exam-style practice questions

1 PDCA is an acronym that is used in continuous improvement. What does it stand for? (2 marks)

....

....

2 A production machine is producing 120 items an hour with a failure rate of 5%. How many rejected items will be produced in one eight-hour shift? (2 marks)

....

....

3 Explain why moving products round the factory between processes is wasteful. (4 marks)

....

....

....

....

Long-answer exam-style practice questions

1 You are a member of the production engineering team in a small engineering firm. You are asked to complete a value stream mapping form for a process. Complete the form, some of which has been filled in to help you. (6 marks)

Step	Activity	Value added?	Type of waste evident
1	Parts are unpacked	Yes	N/A
2	Parts are taken by operative to stores	No	
3	Parts are in stores until needed		Waiting
4	Parts are assembled into final product	Yes	N/A
5	Product put into stores	No	
6	Final product packed and shipped		N/A

Hint

Note that at any stage where the value added is significant, the last column is marked N/A (not applicable). Any wastes should be taken from the TIMWOODS list.

2 Explain how the use of a shadow board can lead to improvements in performance. (6 marks)

3 Explain why a company might implement a process of continuous improvement. (8 marks)

Project and programme management

Recall activity

1 Match the key terms to the definitions.

Term
Minutes
Video call
Documents
Meetings

Definition
Written, printed or electronic records of data or text
The record of what is discussed during a meeting and any resulting actions
Planned occasions when several members of a team come together to discuss a project
Technology that allows meeting participants to see each other in real time

Short-answer exam-style practice questions

1 Rearrange these five stages of a project into the correct order. (4 marks)

Monitoring **Planning** **Implementation** **Evaluation** **Initiation**

....

2 What does FMEA stand for? (1 mark)

....

Long-answer exam-style practice questions

1 You are the manager of a planning department in a small engineering company. The company is becoming very successful and is growing fast, taking on more employees and moving to larger premises across the country. Explain why you might be moving from traditional methods of planning and control to computer-based techniques. (7 marks*)

Plan your own answer

Your answer should cover these areas:

- introduction – definitions of planning and control
- advantages of using computer-based systems
  - efficient
  - data is saved and backed up
  - emails can be sent quickly to different locations
  - documents can be shared between sites
  - everyone works in the same style for important documents
- disadvantages of using computer-based systems
  - expensive to install and provide training
  - all staff need training and access to the system
- conclusion.